Project Management: Survival and Success

Project Management: Survival and Success

Kenneth Lee Petrocelly, PMP

LONDON AND NEW YORK

Published 2020 by River Publishers
River Publishers
Alsbjergvej 10, 9260 Gistrup, Denmark
www.riverpublishers.com

Distributed exclusively by Routledge
4 Park Square, Milton Park, Abingdon, Oxon OX14 4RN
605 Third Avenue, New York, NY 10017, USA

First issued in paperback 2023

Library of Congress Cataloging-in-Publication Data

Names: Petrocelly, K. L. (Kenneth Lee), 1946- author.
Title: Project management : survival and success / Kenneth Lee Petrocelly, PMP.
Description: 1 Edition. | Lilburn, CA : Fairmont Press, Inc., [2018] | Includes bibliographical references and index.
Identifiers: LCCN 2017059938 (print) | LCCN 2018010361 (ebook) | ISBN 9788770222624 (Electronic) | ISBN 0881737933 (alk. paper) | ISBN 0881737941 (electronic) | ISBN 9781138543447 (Taylor & Francis distribution : alk. paper)
Subjects: LCSH: Project management.
Classification: LCC HD69.P75 (ebook) | LCC HD69.P75 P47537 2018 (print) | DDC 658.4/04--dc23

LC record available at https://lccn.loc.gov/2017059938

Project management : survival and success / Kenneth Lee Petrocelly
First published by Fairmont Press in 2018.

Routledge is an imprint of the Taylor & Francis Group, an informa business

Publisher's Note
The publisher has gone to great lengths to ensure the quality of this reprint but points out that some imperfections in the original copies may be apparent.

ISBN 13: 978-87-7022-948-7 (pbk)
ISBN 13: 978-1-138-54344-7 (hbk)
ISBN 13: 9788770222624 (online)
ISBN 13: 978-1-00-315133-3 (ebook master)

Dedication

This book is dedicated to all the people I've worked with and learned from, on and about projects, and it is my way of paying it forward to the next generation of project managers.

Table of Contents

Foreword

However adept I've become as a project manager, it was never my intent to make project management my career objective. Like most other "certified" project managers, my first exposure to the practice came out of the blue, so to speak, when (as a plant operations manager at a hospital) I was tasked to design and construct a parking lot for the hospital staff and its visitors. In the ensuing years, in the position of chief engineer at a power plant, director of facilities at health-care institutions and as a government contract consultant, I found myself called upon to oversee more frequent and increasingly complex projects. This required me to pursue higher levels of knowledge and understanding of the project management profession, until I finally caved and went all out to prepare for and acquire my PMP (Project Management Professional) certification.

The premise of this book is too help the reader avoid many of the pitfalls associated with that means of rising into the ranks, while at the same time providing a tentative path to becoming an accomplished PM.

This book is organized into three sections:

Part I: THE NOVICE—explains the why, who, what, when, where and how of project management, reviews the mechanics of it and describes some of the tools utilized in its practice.

Part II: THE COORDINATOR—gets more into the "people" side of the PM equation, addressing project teams, stakeholder interests, communications and conflict resolution.

Part III: THE PRACTIONER—gets into the meat of the subject, reviewing the PMI (Project Management Institution's PMBOK (Project Management Book of Knowledge), citing the different organizational project management models and suggesting a direction for the reader to pursue to become a competent PM.

Throughout the text, the I share many of my on-the-job project experiences, what I learned from them and how it made subsequent projects easier to lead, survive and succeed.

Preface

If I've learned anything from my many years as a project manager, it's that there is always something more to learn about managing projects. This holds true at every level of project management practice, but it is especially relevant for those company and government managers and supervisors who are tasked by their organizations to perform double duty as project managers. Such practice often produces serious repercussions, either in a form detrimental to an organization's reputation or bottom line, but more often adversely impacting people's careers.

Project Management: Survival and Success addresses the weaknesses and shortcomings of the untrained and/or inexperienced project manager. It views project activity from the perspective of the manager or supervisor who is assigned by their organization to manage a project in addition to his/her normal duties, and explores the four classes of project managers: First, those who are untrained and inexperienced; second, those who are untrained but have some experience; third, those who are formally trained but have no experience; and fourth, those who are formally trained and well experienced.

It also discusses what are considered to be the three major categories of projects: 1) industrial: requiring massive capital investment, and meticulous oversight of finance, progress and quality (as in high-end construction and engineering endeavors); 2) manufacturing: aimed at producing an end product of some consequence or complexity (such as IT software iterations, ships, planes or automobiles); and 3) managerial: projects that arise out of organizational need (building additions and renovations, new equipment installs, department relocations, computerization of operations, landscaping… the list is endless).

This text will un-complicate the project management process and provide direction to ad hoc (sometimes referred to as "accidental") project managers, furthering their understanding and involvement in the profession through referenced examples of actual performed project work. Though the focus of the work deals with the evolution of the incidental project manager (from novice to competent), their fellow project stakeholders (sponsors, customers, team members, et al.) can benefit from the instruction and advice offered herein as well.

Part I — The Novice

Chapter 1

Welcome to My World

BY "MY WORLD" I MEAN everything related to or having to do with project management, from its planning and funding to its rules and tools. Project management is an ever evolving discipline that enables project managers (whether certified or not) to complete projects in a timely and cost effective manner. But this hasn't always been the case. Certainly, projects have been conducted for millennia (think pyramids and the Great Wall of China) but, as a profession, PM has only been around for a relatively short time, getting its start in the 1950s when PM tools and techniques were developed and systematically applied to complete complex engineering projects.

In prior years, projects were managed on an ad-hoc basis, using mostly Gantt charts and other informal techniques and tools, such as the Critical Path Method (CPM) developed by the DuPont and Remington Rand Corporations for managing plant maintenance projects. And the Program Evaluation and Review Technique (PERT), developed by the United States Navy, Lockheed Corporation and Booz Allen Hamilton as part of the Polaris missile submarine program. During that period, project scheduling models were being developed, as was technology for project cost estimating, cost management and engineering economics. (These will all be explained in Chapter 4, The Project Toolbox.)

Since then, with the creation of oversight organizations, the project management discipline has grown in sophistication into what is now a full-blown, highly recognized and valued profession. In 1967 the International Project Management Association (IPMA) was founded in Europe as a federation of several national project management associations, offering a four-level certification program covering technical, contextual, and behavioral competencies. In 1969, the Project Management Institute (PMI) was formed in the United States which publishes "A Guide to the Project Management Body of Knowledge (PMBOK Guide)" and offers multiple certifications as well.

PMI and the PMBOK will be discussed at length in Chapter 3: Getting to Know Project Management and Chapter 10: Understanding the Framework.

BEGINNING THE JOURNEY

As mentioned earlier, before gaining legitimacy as a disciplined profession, projects were managed on an ad-hoc basis. To this day, many organizations still practice what can be termed as impromptu or provisional project management, performed by people sometimes referred to as "accidental" or "incidental" project managers who are bereft of formal PM training. If I've learned anything from my past 40+ years as a project manager, it's that there is always something more to learn about managing projects. This holds true at every level of project management practice, but is especially relevant for those company and government managers and supervisors who are tasked by their organizations to perform double duty as project managers. That practice often produces serious repercussions, either in a form detrimental to an organization's reputation or bottom line, or more often adversely impacting people's careers.

In his play, "Twelfth Night," William Shakespeare declared, "Some are born great, some achieve greatness, and some have greatness thrust upon them." Whereas the nobility of that era may have held such sway, today's project managers do not. They possess no such birthright, rarely achieve great measure from their hard work, and often buckle under the stresses and strains imposed by the projects that are thrust upon them. This is especially true for middle managers who are assigned to projects that they are ill-prepared to undertake due to a lack of training, tools or experience in project management, and/or subject to severe time, budgetary and resource constraints. Project managers fall into one of four categories; 1) those who are untrained and inexperienced, 2) those who are untrained but have some experience, 3) those who are formally trained but have no experience, and 4) those who are formally trained and well experienced. When I started out as a fledgling plant operations manager, my project management expertise fell into the first category. My PM activity was limited to work assigned to me by upper management, requiring me to lead minor projects, in addition to my normal duties. Oftentimes those projects were ill-conceived, unbudgeted, poorly researched for actual need, quickly entered into or just not very well thought through before they were assigned.

MY FIRST PROJECT ENCOUNTER

In the book's foreword, I told you that I was tasked to design and construct a parking lot for the hospital staff and its visitors at the facility where I worked. This was one of the ill-conceived, unbudgeted, poorly researched, quickly entered into and not very well thought through assignments to which I just alluded. At the time, the hospital was expanding its services and had submitted a "Certificate of Need" to the state for increasing the number of beds it was allowed. In anticipation of its approval, the administrator (my boss), recognized that an additional parking capacity would be needed to accommodate the increased flow of visitors and outpatients caused by the expansion. The facility had an organized (finished) parking lot outside of the front entrance for everybody (physicians, department staff, outpatients, vendors and visitors). The administrator's idea was to limit access there to physicians, outpatients and visitors, and restrict staff and vendor parking to the rear of the hospital.

The large asphalted area in the rear of the facility was home to the loading dock and trash compactor and was infrequently used for overflow (helter skelter) parking during special occasions.

In his mind, the conversion would simply be a matter of painting parking spaces onto the existing asphalt lot. Sooo... as the plant operations manager, I was assigned the responsibility to "make it happen." Putting on my managerial "thinking cap," I surmised that there might be a little bit more to it than that, and asked for a week to consider and work through any side issues that might need to be addressed. Thankfully, I was granted that request.

Back in my office, with a fresh cup of coffee in hand, I had mixed emotions about the assignment. As I pondered the work, I felt both exhilaration at the prospect that I would be making a much needed improvement at the medical center and a sense of foreboding regarding the difficulties that the project might pose. I pulled out a tablet and began listing all the things that the front (finished) lot had that the rear (unfinished) lot was missing:

- A logical design (striping and marking)
- Handicap parking spaces
- Sidewalks and ramps
- Parking barriers
- Signage
- Drainage
- Outdoor lighting
- Landscaping
- A schedule for maintenance and repair
- SOPs for sweeping and snow removal

Then I listed the things that the rear lot had that the front lot did not:

- A cardboard box compactor
- A trash dumpster
- The loading dock
- A cage full of medical gas cylinders

Did I say "a little bit more to it?" Were there standards for all of that? Were there construction rules and regulations? Were there municipal (zoning) codes and statutes to contend with? How could I provide for the functioning of the loading dock,

Typical POM Duties	*Comparable PM Responsibilities*
Oversee department operations	Oversee project processes
Monitor service agreements	Monitor service agreements
Write job descriptions	Write job descriptions
Perform employee appraisals	Perform employee appraisals
Maintain blueprint library	Maintain blueprint library
Write equipment operating procedures	Supply operating procedures
Order/stock spare parts/materials	Supply spare parts/materials
Provide employee training	Provide client training
Operation and repair of equipment/systems	Installation of equipment/systems
Identify (tag) system components	Identify (tag) system components
Maintain manufacturer's instructions library	Provide manufacturer's instructions
Track utilities consumption	Provide temporary utilities
Interpretation of test results	Documentation of test results
Perform in-house renovations	Conduct renovations projects
Committee memberships	Attendance at meetings
Regulatory compliance	Regulatory compliance
Fire protection	Interim life safety measures
Physical plant systems evaluation	Physical plant systems installations
Energy auditing	System commissioning
Quality Assurance Programs	Quality Management Plan (CQM)
Emergency contingency planning	Project work-arounds
Preventive maintenance efforts	Provide PM procedures
Life safety code compliance	Life safety code compliance
Exterior (grounds) maintenance	Landscaping
Hazardous materials management	Hazardous materials management

compactor, dumpster and medical gas cylinders without interfering with the staff's way of travel into and out of the hospital?

COMPARING OPERATIONS TO PROJECT MANAGEMENT

I was a good operations manager, but I questioned whether or not I could successfully take on the mantle of project manager. At this juncture of my career I had managed myriad processes, but never lead a project. The thought came to me that they must be kindred spirits of sorts; so I went about the task of determining what their similarities were and if they had any differences. To my surprise, they had a significant number of both. Most work effort generally involves either operations or projects. Both are performed by people, constrained by limited resources and are planned, executed, and controlled. Each has a different set of objectives. Operations entail permanent endeavors that produce repetitive outputs, such as in factory production, software service management, building maintenance, equipment repair and accounting transactions; whereas projects are temporary and unique, have a definite beginning and a definite end, and are usually undertaken to create a product or service. Projects require project management expertise whereas operations require business process management or operations management abilities.

Organizationally, operations managers differ from project managers. For instance, operations managers may have many different skills, while the project manager's skills may be specialized (such as in the information technology arena). The ops manager likely has a long-standing personnel organization, while the PM's team members disperse once the project has been completed. Whereas the operations manager has a daily routine whose work is repetitive, the PM's role ends with the project. Too, the ops manager has an annual planning cycle, while the project manager must contend with time, cost and scope constraints.

THINKING LIKE A PROJECT MANAGER?

That last part got my full attention. I started speculating as to what the scope of the project would be, so that I could estimate how much it would cost and how long it would take to complete. It was a daunting task. I first made inquiries into what state and local requirements there were for creation of surface parking lots, and learned what was available regarding the engineering aspects of the work. While pouring over the information, it dawned on me that I had the requisite knowledge I needed, at hand, the whole time. Though I wasn't working at the facility when the front lot was constructed, it was designed and completed by the appropriate professionals (architects, engineers and tradesmen), in accordance with regulatory requirements, and it passed inspection before being put into service. All I had to do was replicate, in the rear of the building, what they had constructed at the front (with some modifications). With that in mind, I began to assemble my scope of work for the project. The following were my considerations.

Parking Geometrics

There are a wide range of automobiles on the road from the "minis" and "micros," to light trucks and sport utility vehicles, to conversion vans and multi-passenger pick-up trucks, and they can vary by region. At the hospital, parking spaces served five groups of users: patients, visitors, physicians, vendors and staff. As before mentioned, the first three would be utilizing the front lot, and the latter two would be relegated to the rear lot.

Space Width and Length

The critical elements of parking space dimensions are stall width to vehicle width, and the ease of maneuvering into and out of the parking space. The patient and visitor lot had 9′-0″ wide by 18′-0″ long 90-degree parking stalls throughout the front lot. None of the stalls was reduced in size for "small" or "compact cars." [See Figure 1-1.]

HOSPITAL FRONT

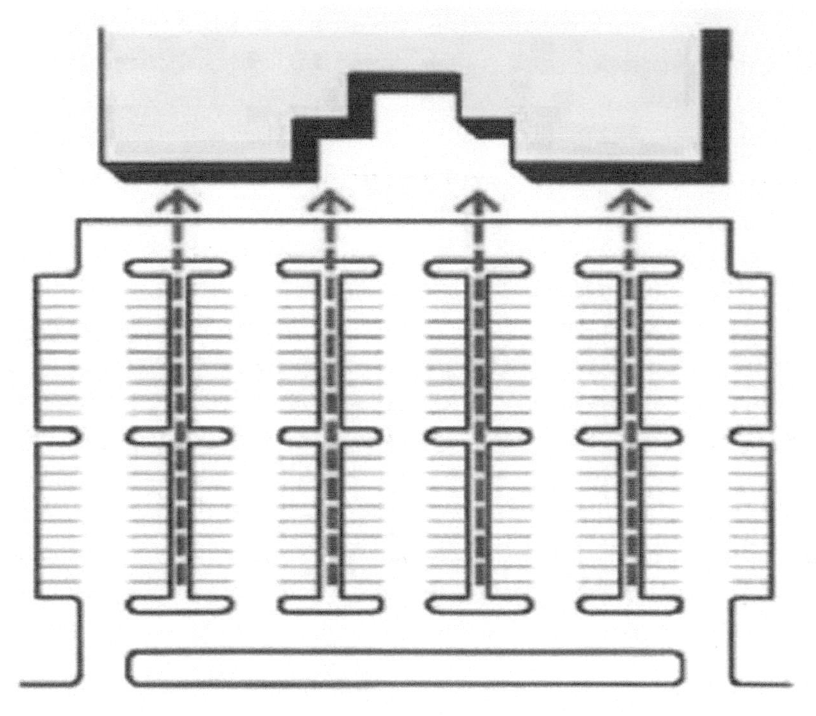

Figure 1a: (Layout)

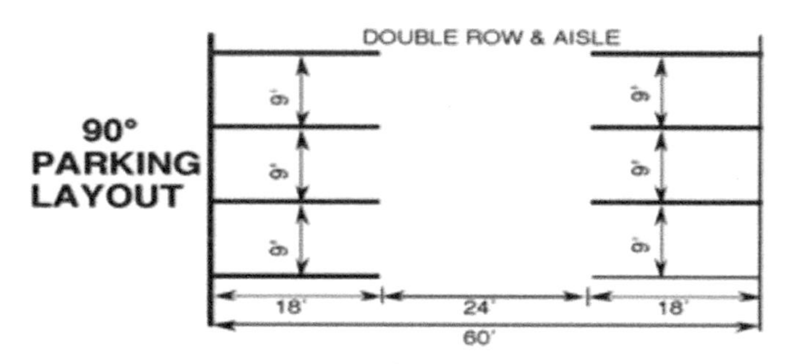

Figure 1b: (Geometrics)

Figure 1-1. Front Parking Lot

Due to the need for a work-around in the rear lot, to accommodate the service areas and items, the layout of the parking spaces needed to be modified. After addressing how the respiratory therapy staff could access their medical gas cylinders, how the housekeeping personnel could get to and from the compactor and dumpsters, how the receiving department could take deliveries at the loading dock and how vendors could perform their functions (all without interfering with the flow of pedestrian traffic from the new lot), we brainstormed the best configuration of the parking spaces that would yield the most parking stalls with the least travel-flow problems. How's that? Yes, I said we. At that juncture, I realized that the work involved more than just assigning people tasks to accomplish the work; project work required a team effort. [See Figure 1-2.]

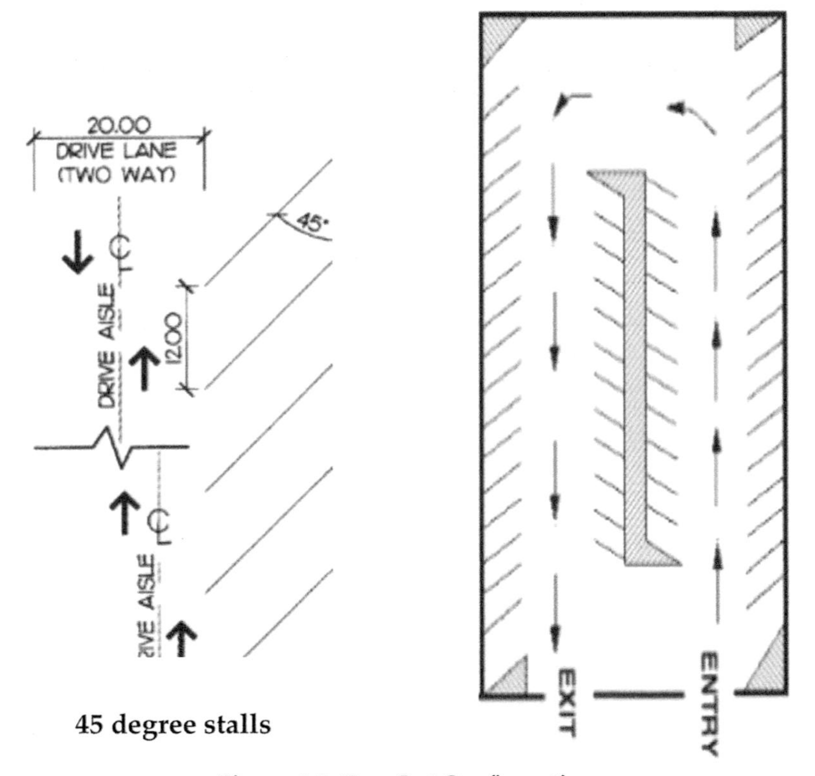

Figure 1-2. Rear Lot Configuration

Circulation

One-way drive aisles provide better traffic flow than two-way drive aisles because it is easier for users to enter and exit parking spaces, and there is less circulation conflict. Two-way traffic offers wider drive aisles, making for a safer environment for pedestrians and passing vehicles. [See Figure 1-2.] Generally speaking, it is preferred not to combine one-way and two-way traffic flows within a single facility, but due to the problems confronting us in the rear lot, we had little choice.

Surface Lot Design

Parking lots should be designed to be convenient and have an aesthetic appearance. Entrance and exit points should be limited in number to reduce internal conflict and configured to avoid dead end parking conditions. Walkways should be designed to safely separate pedestrians from vehicular traffic. To accomplish that, along the pedestrian path between the lot and the rear entrance to the hospital, we designed in trees and bushes, pathway lighting, and amenities such as trash receptacles and park benches.

Pavement, Water Management and Landscaping

My team (our engineering and maintenance personnel) agreed that we would need to hire a contractor to both ensure the monolithic integrity of the existing asphalt (in conformance with local DOT standards) and the proper degree of slope for adequate draining of storm-water, to comply with EPA storm-water regulations. Water quantity and quality measures must be implemented as part of the lot's overall drainage strategy. (Surface lots should be crowned a minimum of 2 percent toward drain inlets, catch basins, or curb inlets).

Landscaping provides an attractive and user friendly site feature. Vegetation can also effectively reduce runoff. A significant amount of storm-water can evaporate from beds of flowers, shrubs and trees. Landscape materials will be used to define the lot boundaries and to screen medical gas cylinder storage and utility areas, and careful consideration will be given to avoiding driver line-of-sight and security problems.

Lighting

For the sake of standardization, the same type of outdoor lighting used in the front lot will be installed in the rear lot. A contract will need to be entered into with the current company that maintains and originally installed those lights. The light standards (and illumination requirements) will comply with the dictates of the National Electrical Code and other applicable codes for exterior lighting. Light standards will not be located inside the parking field, but will be located outside of the parking lot perimeter and inside the lot's island curbs.

Handicap Parking

Author's note: Although the hospital (as a consequence of the work they performed) did provide what were then referred to as "handicap" parking spaces, it was operating under the dictates of The Rehabilitation Act of 1973, built off of the Civil Rights Act of 1964 and The Architectural Barriers Act (ABA, 1968). The Equal Opportunity for Individuals with Disabilities Act (better known as the Americans with Disabilities Act, or ADA) was signed into law and became effective on July 26, 1990 as 42 U.S.C., Chapter 136; Section 12101. The Rehabilitation Act of 1973 prohibited discrimination on the basis of disability in programs conducted by federal agencies and those receiving federal financial assistance. ABA and the Rehabilitation Act essentially extended certain civil rights to people with disabilities. Like its predecessors, ADA is a civil rights law. And while not specific to transportation, it is difficult to find a more far-reaching transportation policy for the disabled within or outside of the U.S. Unlike its predecessors, ADA applicability is not tied to whether federal funding was involved in the construction of the facility; hence, it applies broadly to any facility open to public use.

Signage

All signage will be homogenous with existing facility signage and marking requirements.

THE REALITY CHECK

Taking into account a project's time, cost and scope constraints, (none of that had been discussed earlier when the project was first assigned), the more we thought through it, the more the project seemed to grow (in all three respects). Bottom line: It was time to lay all the cards on the table. After an initial "shock-and-awe" meeting with the administrator, he scheduled us to make a presentation to the board of directors to sell them on the design and the need for those third-party contracts. They were receptive to our plan (so much for the scope aspect) and we got their permission to move forward with the project. Nothing was to come out of the plant operations department budget, as all expenses would be paid for out of the hospital's contingency funds (but we still put together a budget to track and control the expenditures). As regards the time constraint, we created a schedule that we estimated would allow the project to be completed prior to approval of the certificate of need and setting up of the expanded operations.

LESSONS LEARNED

Though we succeeded in completing the project in good time and received a lot of praise for its appearance and functionality, I felt confident but not yet competent as a project manager. As a fledgling operations manager, I knew that in my position I would be called upon to lead many more "incidental" project efforts in the future. Before looking to pursue formal training in the profession, I opted to gain some insight from this first experience by throwing a victory party (pizza and sodas) for my crew. The party doubled as a review of what we learned during the course of the project. Bite by bite we recalled the projects activities and listed the items that most stood out. Though not a complete list of the takeaways (in no particular order), the following is the gist of what the group came up with:

- It was advantageous to thoroughly think through the needs of the project.
- Working off of a scheduled plan was paramount.
- It took a true "team" effort to accomplish our goals.
- Frequent communication with upper management was important.
- Contract negotiation skills were critical for vendor involvement.
- Coordination of our in-house and vendor efforts was crucial.
- Timing of work schedules was necessary keep the project from stalling.
- Operations and projects have similarities and differences.
- We became well versed in rules and regulations.
- It was necessary to utilize the talents of other department personnel.
- Alternating project work with day-to-day duties was burdensome.
- We needed to be better educated on the three primary project constraints.

I used that list to put together a stratagem for my department and me to be better prepared for the next "surprise assignment." I'll share that with you in the next chapter.

Chapter 2

Taking on the Challenge

IT WASN'T LONG AGO that plant operations and facility managers considered an infrequent construction project an enjoyable divergence from the humdrum of their daily routines. But as capital building funds dry up and corporate operating budgets are cut to the quick, our involvement in such activities at once becomes more commonplace and complex. Coupled with the dwindling money supplies, our aging physical plants have forced us to look inward for more creative means of accommodating our organizations' needs for change. Accordingly, we've had to change too. Once only an avocation, project management has transformed us from caretakers to code interpreters, imparting on us the duty to build (or at least cause to be built) the very facilities and structures we are charged with maintaining. This transformation is triggered by a cadre of rationalizations, including:

- code violations and new regulatory mandates
- provisioning of amenities for customers and staff
- expanded markets and program aberrations
- aesthetic and environmental improvements
- anticipation of new services and/or improved productivity
- increased demands on existing resources
- changes in the way existing space is utilized
- overloading and obsolescence of systems
- upgrading to state of the art
- long range corporate planning

THE CONSTRUCTION DECISION

The trouble with project work is that no two jobs are ever alike. Like ulcers, they come in all shapes and sizes; some you must give serious consideration to while others are easily remedied. By its nature, most project work is a one-shot deal that requires staffing only until the work has been completed. There are times when you'll have projects scheduled back-to-back for months on end, then have slack periods upwards of a year where no major work needs to be performed. How should you staff your department to accommodate these inconsistencies? Do you hire full-time personnel in anticipation of the work? What happens when the project is canceled? Do you hire a full-time crew on a temporary basis? Good people are hard to come by unless you offer them more secure employment. Do you use contractors? Depending on the scope of the project this may become necessary, but bear in mind you'll pay a premium for the luxury of hiring outside help. Can you use existing staff personnel? The truth is, your staff should be capable of handling all but the most involved projects confronting your department.

To a lesser degree, the knowledge needed for maintaining your building and systems is essentially the same as the knowledge required to construct and install them. Constructing a building may entail the need for employing a contractor, but renovations and small additions can be tackled in-house. The installation of a complex system may warrant the hiring of outside help, but the modification of that system may be accomplishable by your personnel. When department personnel double as construction crews you save on the cost of premium wages paid to outside vendors, you avoid the hiring and lay-off process, and your people take pride in accomplishing the work themselves. The choice to use them in this way is yours, barring any restrictions that may be imposed upon you by a bargaining agreement. Just remember to change their job descriptions before attempting to change their jobs.

Whatever triggers your need to doff your maintenance mantle

and don your contractor's cap, rest assured you'll have more questions than answers from the start. Initially, determinations must be made regarding permitting, drawing approvals, inspection requirements and whether or not to employ a registered architect or professional engineer. Subsequently, a decision must be made as to the extent of your operations involvement in the project. Concurrently, a multitude of inquiries must be fielded and laid to rest, from demolition through final acceptance. Such speculations can be pure joy or torture depending on how you approach them. If you have anywhere near the threshold for pain that I do, it will behoove you to keep things simple. One way to do this is by constructing a decision tree to help you think through the process. For instance, after establishing the scope and magnitude of the work, categorize the project as either a (type 1) rudimentary renovation that requires neither formal architectural drawings nor engineering to complete, such as a simple office relocation; a more complex scheme (type 2) which calls for limited A and E assistance, such as the construction of a small outbuilding to house your grounds keeping equipment; or a more comprehensive design (type 3) that mandates the need for extensive evaluations to be performed by experts outside of your organization, such as the creation of a new service wing which will tap into your building's existing utilities. At this point, a decision can be made as to what extent you might use your people on the project. You can then proceed through the next step to decide on which type of contract to let; i.e., design-bid-build, design/build, construction management, owner/builder and so on. (See Figure 2-1.)

KEY CONSIDERATIONS

Regardless if the project you're confronted with is a renovation or new construction, certain aspects common to both should be considered during the course of its completion. Aside from the more mundane items such as funding, site-selection and the bidding process, there are some common sense issues that should

Figure 2-1. Blueprints are the Project's Roadmaps

be addressed. Though readily apparent to those who work within the confines of a facility, many issues aren't recognizable by persons foreign to specific operations. Such concerns deal with items like the interdependency of departments, the remaining capacity of buildings systems, normal operating hours and customary traffic patterns. Other significant matters include:

Concurrent Occupancy

What extraordinary measures need to be taken to assure operational continuity during construction?

Construction Materials, Fire Ratings

Have the flame spread ratings been specified on all materials to meet code?

Extraordinary Measures

Will any special equipment need to be employed such as hiring of a crane to set roof-mounted units?

Materials Handling

Have you allocated storage space for building materials and

designated one elevator as "freight only?"

Fire Safety and Security Measures

Have you arranged for a fire watch and or security guard/ alarm service?

Tapping of Existing Systems

Were load studies performed prior to accessing the old systems?

Conceptualization

Has the project been thought through to the end?

Input from End Users

Did anyone think to ask for the opinions of the people who will ultimately live with the changes?

Outside vs. In-House Construction

Was a determination made as to the costs and implications of each?

Salvage of Existing Components

Can any portion of the existing structure or equipment be reused?

Temporary Services

Were arrangements made to assure against sudden accidental shutdowns?

Standardization

Standardize, standardize, standardize!

Functional Adjacencies

Will the new operation be compatible with the operation abutting it?

Figure 2-2. Construction Cost Spreadsheet

Work Item	Vendor	Labor	Equipment	Materials	Subcontr.	Subtotal	Markup %	Markup	Total
Permits/Fees	City of Los Angeles				$1,500.00	$1,500.00		$0.00	$1,500.00
Excavation		$6,000.00	$8,000.00	$500.00		$14,500.00	15.00%	$2,175.00	$16,675.00
Utilities		$3,500.00	$2,500.00	$2,750.00	$1,000.00	$9,750.00	15.00%	$1,462.50	$11,212.50
Water Well						$0.00		$0.00	$0.00
Septic Tank						$0.00		$0.00	$0.00
Foundation	Connie's Concrete				$3,500.00	$3,500.00	5.00%	$175.00	$3,675.00
Concrete Flatwork	Connie's Concrete				$1,900.00	$1,900.00	5.00%	$95.00	$1,995.00
Framing		$3,500.00	$1,500.00	$9,000.00		$14,000.00	15.00%	$2,100.00	$16,100.00
Roofing	Robert's Roofing				$3,500.00	$3,500.00	5.00%	$175.00	$3,675.00
Windows/Ext.Doors	Wally's Windows				$8,000.00	$8,000.00	5.00%	$400.00	$8,400.00
Garage Door	Gary's Garage Doors				$2,250.00	$2,250.00	5.00%	$112.50	$2,362.50
Siding						$0.00		$0.00	$0.00
Electrical	Ernie's Electric				$18,500.00	$18,500.00	5.00%	$925.00	$19,425.00
Plumbing	Mac's Mechanical				$16,500.00	$16,500.00	5.00%	$825.00	$17,325.00
HVAC	Mac's Mechanical				$23,000.00	$23,000.00	5.00%	$1,150.00	$24,150.00
Insulation		$3,500.00		$1,000.00		$4,500.00		$0.00	$4,500.00
Masonry	Mason's Masonry				$14,500.00	$14,500.00	5.00%	$725.00	$15,225.00
Drywall	Doug's Drywall				$12,500.00	$12,500.00	5.00%	$625.00	$13,125.00
Interior Trim	Doug's Drywall				$9,000.00	$9,000.00	5.00%	$450.00	$9,450.00
Painting	Paul's Painting				$13,500.00	$13,500.00	5.00%	$675.00	$14,175.00
Floor Coverings	Carl's Carpets				$16,500.00	$16,500.00	5.00%	$825.00	$17,325.00
Cabinets	Ken's Kabinets				$22,500.00	$22,500.00	5.00%	$1,125.00	$23,625.00
Appliances	Abby's Appliances	$2,500.00		$11,500.00		$14,000.00	15.00%	$2,100.00	$16,100.00
Landscaping	Sonny's Sodding				$2,750.00	$2,750.00	5.00%	$137.50	$2,887.50
Overhead Costs		$10,000.00				$10,000.00	20.00%	$2,000.00	$12,000.00
Other						$0.00		$0.00	$0.00
						$0.00		$0.00	$0.00
						$0.00		$0.00	$0.00
						$0.00		$0.00	$0.00
						$0.00		$0.00	$0.00
						$0.00		$0.00	$0.00
TOTALS						$236,650.00	7.71%	$18,257.50	$254,907.50

Project Management

Do you have the time to devote to the project or should you supplement your capabilities?

Interior Design

Is there an approval system in place to forego the displeasure syndrome?

Meetings/Minutes

Are timely meetings held between the project principles and accurate minutes recorded?

Attic Shock

Have sufficient overages been built into the materials budget to assure a maintenance stock?

Cost Overruns

Are expenditures strictly monitored, controlled and regularly reported to upper management? (See Figure 2-2.)

If, after considering all of the fore mentioned idiosyncrasies, you are determined to accomplish the feat utilizing only in-house talent, don't forget to take into account the extraordinary costs associated with such an endeavor; both monetary and operational. Even the smallest of jobs will require significant outlays of cash for tooling up and every hour devoted to new construction is one hour less spent on maintaining your existing facility.

THE PLANNING PROCESS

The diversity of project work precludes the consistent use of exacting plans. Whereas all project plans incorporate the same basic elements in their development, individual projects dictate the order and intensity of their application. Here is a checklist of some of those elements to consider:

- feasibility studies

Figure 2-3. Construction Handbooks and Manuals

- pre-plan discussions with affected departments
- selection of project team members
- hiring of sub-contractors
- negotiation of contracts
- narration of bid requests
- specification of materials
- tools and techniques to utilize
- acquisition of required permits
- site preparation
- acceptability of change orders
- compliance with rules and regulations
- interruptions of operations

- completion dates and deadlines
- ordering of materials
- getting required approvals
- cost estimating
- scheduling of progress meetings
- setting up a communication network
- labor and material expenditures
- reuse of existing materials and structures
- inspections and testing
- maintenance of forms and records
- distribution of drawings and specs
- insurance and bonding requirements
- demolition of existing structures
- completion of punch lists
- phasing of project levels
- receiving and storing materials
- contingency planning
- using A&E firms and consultants
- instruction and scheduling of personnel
- temporary bathroom facilities
- coordination of utility hook-ups
- protection from the weather
- isolating the job site from the public
- designation of lead personnel
- design constraints and scope creep
- mapping out of the mechanicals
- recourse for bad workmanship
- tolerability for substitutions
- securing the work site
- job cleanup and safety

ADMINISTERING THE WORK

How to document your projects is a singular decision. If you haven't got an inkling as to where to begin, here are a few ideas. Construction projects can generate an inordinate amount of paperwork. Before beginning any project, make certain you have sufficient file space to control it. The number and scope of the projects you take on will dictate what space is needed. It is advisable to devote at least an entire file cabinet to project documents and its drawers should separate projects by status as pending, in progress or completed. Much of the supporting documentation associated with project work, such as purchase orders and invoices, is redundant to your normal operation and as such should remain there but be copied to the project folder for purposes of reference. Other information normally found in a project file includes:

- planning scenarios and photographs
- vendor contracts, bonds and insurance certificates
- building permits
- correspondence
- minutes from project meetings
- architectural and engineering renderings
- work schedules
- progress reports
- guarantees and warranties
- change orders and punch lists
- equipment manuals
- material specifications
- approvals

Projects can be likened to football games in that they both require game plans to insure their success. In the project game, you are the general manager, your supervisors are the coaches, and your personnel are the players. As usual, the owner is relegated to picking up the tab. If a project can be equated to a game, then all the projects completed in a year constitute a season. How well you conceive and implement your individual game plans determines whether you end each season as a contender or as an also ran. Just as each opposing team poses a different challenge causing you to alter your game plan, each project presents special problems, which cause you to modify your work plan for completing it. Whereas all teams have basic plans, only those whose plans are flexible enough to meet those challenges will wax victorious.

SALVAGE OPERATIONS

Most project work performed by your department will entail the modification of existing structures. New is nice, but it's not always the answer. There is intrinsic value in reworking portions of an old structure resulting in lower manpower requirements, less use of materials and enhancement of the structure's original character. Thought should be given to repairing and/or reusing:

- lock and passage sets
- door butts, panic hardware and closers
- exterior windows
- suspended ceiling grids
- plumbing fixtures
- built-in cabinetry
- railings and banisters
- decorative stone
- metal ducting and plenums
- decorative molding
- air registers and diffusers
- fluorescent light fixtures

This technique must be administered selectively, for as much as you can gain from reusing some items, you can lose on others. Never reuse old electrical wiring and devices. The ashes that fall on the floor may be yours. Never reuse old wood supports. If you want to get to the basement quickly, use the elevator. Never reuse old sewage piping. The idea stinks; if you catch my drift. A good rule of thumb is to reuse only those things you will see when the job is finished. Let common sense be your guide.

WORKING WITH CONTRACTORS

Hooking up with a contractor is very much like making friends in the military during times of war. Generally, you're only together for a short hitch, but it often seems like a lifetime. You end up knowing each other's darkest secrets and innermost feelings, hoping the information doesn't somehow get around. You wonder if one or both of you will come out of the miserable experience alive. This is one relationship you should give serious consideration to before entering into it; a divorce midway through this marriage can literally wipe you out financially. Don't think about signing on with someone until you've answered these questions:

- Did you get at least three bids?
- Were the bidders bidding on identical plans?
- How long have they been in business?
- How reputable are they?
- Have they completed comparable projects?
- Did you get several references from completed projects?
- Are they experienced in this type of construction?
- How much time do they spend on their work sites?
- Are they accessible for discussion?
- How open are their suggestions and requests for change?

- How do their prices compare to their peers'?
- What are their markups for the job?
- Are their companies solvent?
- What kinds of problems have they experienced?
- Do they keep good records?
- What was your overall impression?
- Can they give your project their personal attention?
- Did you visit their headquarters or work lot?
- What insurances do they carry (in what amounts)?
- What is the general condition of their equipment?
- Did they provide you resumes of their employees?
- Do they use union or non-union personnel?
- Have they had labor problems?
- What sub-contractors do they use?
- How long will it take them to complete their work?

GETTING WHAT YOU PAY FOR

Once you are sure that you've checked everything out—that you feel comfortable with the contractor and their design works for you, and after you've determined that you can live within the budgetary constraints—before signing the agreement, make certain that you thoroughly review it (especially the small print). The contract should state who the parties to the contract are and state the consideration. Both the rights of the owner and the contractor should be spelled out and any drawings or documents, and specifications should be included. A section of the agreement should specify material quality and explain what recourse the owner has for poor workmanship or missed deadlines. It should address insurance types and amounts, which codes will be adhered to and who will be responsible for taxes and fees. The contract should also list what schedules and reports to expect and how disputes that arise will be resolved. Other items that it should cover are guaranties and warranties, temporary utilities, site monitoring, barriers, signage and clean-up—just to name a few.

Chapter 3

Getting to Know Project Management

NEEDLESS TO SAY, that first project turned out to be a real eye opener, prompting me to get better acquainted with the profession. The foregoing chapter was a compendium of "early project logic" created to provide the novice PM with a scheme for dealing with ad hoc project assignments, in lieu of a more formal, sophisticated stratagem. Not that it was without merit. In point of fact, most of the subject matter it details pays dividends when incorporated into a structured project management plan. Before acquiring my PMP (Project Management Professional) certification, that tact served me well over the years to successfully complete myriad projects, including:

IN-HOUSE CONSTRUCTION—Design and installation of a 4,500-square-foot family practice residency. Fabrication of a $4 million women's obstetrical pavilion and a $3 million laboratory expansion.

CONSTRUCTION—Demolition of a 7-story hospital building and acquisition of land parcels and permits to build three free-standing medical office buildings (MOB).

EQUIPMENT INSTALLATIONS—Oversaw myriad UPS (uninterrupted power supply)—emergency backup power systems and equipment, including the purchase and installation of a 1200kVA (autonomous) emergency power diesel generator, service entrances for mobile CT (computer tomography) scanners, lithotripters and MRI (magnetic resonance imaging) systems.

IT SOFTWARE INSTALLATIONS—Oversight of several computerized maintenance management systems (CMMS) software implementations (Maximo, PUMMP, Hems).

MANAGERIAL—Performance of several department relocations; post occupancy evaluations (POE); condominium reserve studies and an airport energy conservation program. Wrote capital planning and project management manual (450 pages) and real property administrative and management manual (350 pages); along with several safety and standard operation procedure (SOP) manuals.

Once trained and certified, I was able to take on larger, more complex projects. Armed with a greater comprehension of the profession, I was able to enter into the government contract arena. I'll discuss this more at length in a later chapter, but first, let me share with you what it took to get me there.

WHAT IS PROJECT MANAGEMENT?

Author's note: In Chapter 1, I mentioned the founding of project management certifying organizations like the International Project Management Association (IPMA) and Project Management Institute (PMI). Both are global organizations which were launched to promote the project management profession. IPMA is UK based and PMI is the de-facto standard for the USA. It is the PMI organization and PMP certification that I will be focusing on in this work.

In the project management profession, it goes without saying that to be a project manager, you must have a clear understanding as to what project management is about. The primary challenge of project management is to achieve all of the project goals within the given constraints. A project is a temporary endeavor undertaken to create a unique product, service or result; and a project manager is the person responsible for leading a project from its inception through its closure, through the management of the people, resources and scope of the project. PMI defines project management as the discipline of initiating, planning, executing, controlling and closing the work of a team, to achieve specific goals and meet specific success criteria. To do that requires the application of knowledge, skills, tools, and techniques of project activities to meet the project requirements. PMI global standards provide widely accepted guidelines, rules and characteristics for project, program and portfolio management. When consistently applied, they help project managers and their organizations achieve professional excellence.

One of those standards is the PMBOK (an acronym for) Project Management Book of Knowledge. The PMBOK project management guide is the pre-eminent global standard for project management, providing project professionals with the fundamental practices needed to achieve organizational results and excellence in the practice of project management. It is an internationally recognized standard published by the Project Management Institute (PMI), which houses the entire collection of processes, best practices, terminologies, and guidelines that are accepted within the project management industry. The two main drivers of the guide detail the five basic process groups and the ten knowledge areas that are typical of almost all projects.

The five process groups (in order) are:

- *Initiating*: Defines the project and authorizes its start or new phase.
- *Planning*: Establishes the scope of the project and its objectives.
- *Executing*: Completes the work defined in the project management plan.
- *Monitoring and Controlling*: Tracks, reviews and regulates the progress and performance of the project.
- *Closing*: Finalizes all process activities to close the project or phase.

The ten knowledge areas are...

- *Project Integration Management*: Identifies, defines, and coordinates the various processes and activities within the project management process groups.
- *Project Scope Management*: Ensures that the project includes all the work required to successfully complete the project.
- *Project Time Management*: Manages the timely completion of the project.
- *Project Cost Management*: Estimates, manages, and controls costs to complete the project within the approved budget.
- *Project Quality Management*: Determines policies, objectives, and responsibilities that will satisfy the needs of the project.
- *Project Human Resource Management*: Organizes, manages, and leads the project team.
- *Project Communications Management*: Ensures appropriate creation, distribution, storage, management, control, monitoring, and disposition of project information.
- *Project Risk Management*: Plans, identifies, analyses and controls project risks.
- *Project Procurement Management*: Acquires products and services, needed from outside the project team.
- *Project Stakeholder Management*: Identifies all people or organizations impacted by the project.

The PMBOK guide provides the fundamentals of project management, irrespective of the project type. An in-depth summary of the guide's best practices, tools and techniques, guidelines, and terminologies will be covered in Chapter 10.

TYPES OF PROJECTS

As can be seen by the list of projects at the beginning of the chapter, project management takes many forms. Depending on who is informing you on the subject, there are any number of types, classes or categories of projects. Some project types are based on the product they produce, such as:

- *Construction*: constructing a parking lot or garage
- *In-House Construction*: re-configuring a department
- *Software Development*: new program or iteration
- *Design*: architectural or engineering plans
- *Administrative*: installing a new program or system
- *System Installation*: a steam generation system
- *Department Relocation*: a move to a lower floor
- *Maintenance of Process*: emergency power back up
- *Remodeling*: re-doing a kitchen or bathroom
- *Product Development*: a new appliance or device
- *Managerial*: writing an SOP or safety manual
- *Research*: an energy conservation program

I personally hold that there are three major categories of projects:

1. *Industrial*—requiring massive capital outlays, and meticulous oversight of finance, progress and quality (such as in high-end construction and engineering endeavors).
2. *Manufacturing*—aimed at producing an end product of some consequence or complexity (such as an IT software iteration, plane, train or automobile).
3. *Managerial*—projects that arise out of organizational need (for example, building additions and renovations, new equipment installs, department relocations, computerization, landscaping.) The list is endless.

A PROJECT'S PRINCIPALS

The four individuals or groups of people that comprise the authority and oversight of most projects (the key players) are the sponsor, the project manager, project team members and stakeholders.

The *sponsor* is usually the project authority and initiates most project activity. They provide the resources needed to complete projects and ensure that their organization's goals and objectives are met. Their involvement in the actual project activities is restricted to motivating and supporting the project team.

The *project manager* manages the project as the leader of the project team and is responsible for the completion of the project and ensuring that the organization's goals and objectives are achieved. Their role calls for them to provide a plan of action for and ensure the timely progress of the project, motivate their team and communicate the project's status and problems with the sponsor, their team members and the project's stakeholders.

The *key team members* collaborate with their fellow members to aid the project manager by contributing their expertise and completing their assigned work. They help the project manager determine the feasibility of the project, input into its planning, and help ensure the timely and cost effective completion of the project. The focal point of the remaining team members (the worker bees) is the completion of one or more tasks assigned to them via the work breakdown structure (WBS) [more on that in the next chapter].

The *stakeholders* are comprised of any remaining persons or group having a vested interest (stake) in the project—in essence, anyone who can affect or be affected by the project. They could be contractors and sub-contractors hired to do the work, suppliers of materials, customers or clients, regulatory bodies or simply anyone having an interest in the project. Contractors and sub-contractors are responsible for providing services per their contractual agreements. Suppliers are responsible for providing the resources for a project, including materials, products and services. Customers and clients often influence the project manager's decisions by changing the objectives or direction of the project. And regulatory bodies can be disruptive agents as rules and laws are changed or enacted.

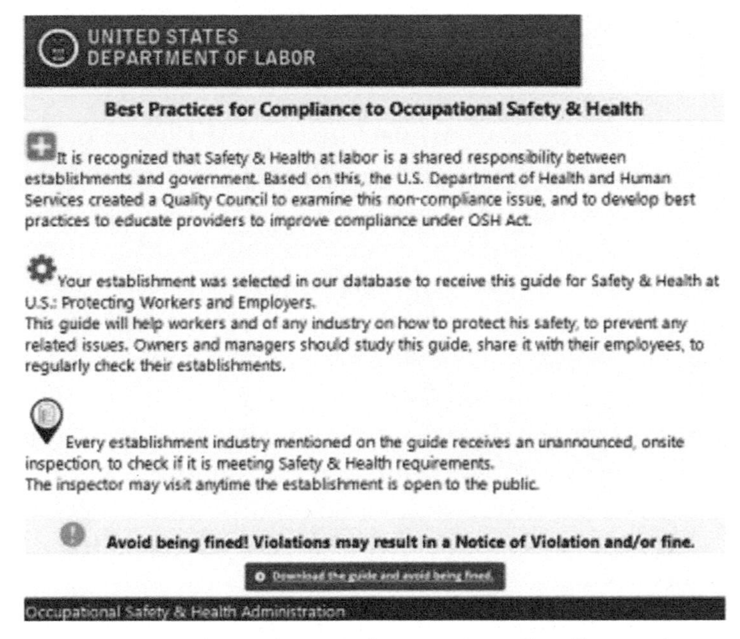

UNITED STATES
DEPARTMENT OF LABOR

Best Practices for Compliance to Occupational Safety & Health

It is recognized that Safety & Health at labor is a shared responsibility between establishments and government. Based on this, the U.S. Department of Health and Human Services created a Quality Council to examine this non-compliance issue, and to develop best practices to educate providers to improve compliance under OSH Act.

Your establishment was selected in our database to receive this guide for Safety & Health at U.S.: Protecting Workers and Employers.
This guide will help workers and of any industry on how to protect his safety, to prevent any related issues. Owners and managers should study this guide, share it with their employees, to regularly check their establishments.

Every establishment industry mentioned on the guide receives an unannounced, onsite inspection, to check if it is meeting Safety & Health requirements.
The inspector may visit anytime the establishment is open to the public.

Avoid being fined! Violations may result in a Notice of Violation and/or fine.

Download the guide and avoid being fined.

Occupational Safety & Health Administration

Figure 3-1. Rules and regulations profoundly affect projects.

POSSIBLE PROJECT OUTCOMES

There are many different situations that govern a project. As you know, every project has project goals and objectives and is carried out under certain constraints; the major ones being schedule (time), cost (budget) and scope. The manipulation of those three constraints (commonly referred to as "the triple constraint or the project management triangle") is the main reason behind the success or failure of many projects. The schedule determines if it will meet the time commitments of the project; the budget determines if it will meet its cost estimates, and the scope determines the specific goals, deliverables and tasks that define the project boundaries. The triple constraint components, cost, time and scope, are related in such a manner that if one changes, then one or both of the remaining two must also change, in a defined and predictable way. (See Figure 3-2.) Project management has the tools necessary to manage and balance their interaction and ensure that only the proper levels of resources are expended. In fact, there are three knowledge areas (project scope management, project time management, and project cost management, that deal with the three (3) triple constraint factors in the project management body of knowledge (PMBOK) guide; that will be covered in Chapter 10.

Besides the *measureable* constraints imposed by scope, cost and time, project managers are confronted with countless other success criteria such as product quality, meeting safety standards, achieving the project's purpose and providing sustainable reliability. To a large extent, project success is in the mind of the beholder. The project manager may consider it a success because the changes to the original plan were kept to a minimum. The sponsor may say it was because it finished on time and under budget. The client (customer) might say it was their satisfaction with the end product. Projects succeed and fail for myriad reasons, but most can be traced back to how supportive the sponsor (usually executive management) was, how clear the requirement statements were (and if they were appropriately conveyed), how realistic expectations were (by all of the stakeholders), and whether or not the project was properly planned. When undertaking any venture, the key to all successful projects is the formulation and carry-out of a well thought through and worthwhile project plan.

Projects can end in four ways: 1) successfully—implying that all parameters were met and that everyone was satisfied with the result, 2) compromised—hinting that one or more aspects of the project didn't meet muster, 3) failed—indicating that the project ended with major problems left unresolved, and 4) unfinished—connoting that it was canceled either before or after its start. Compromised projects come about for a variety of reasons such as from incomplete requirements,

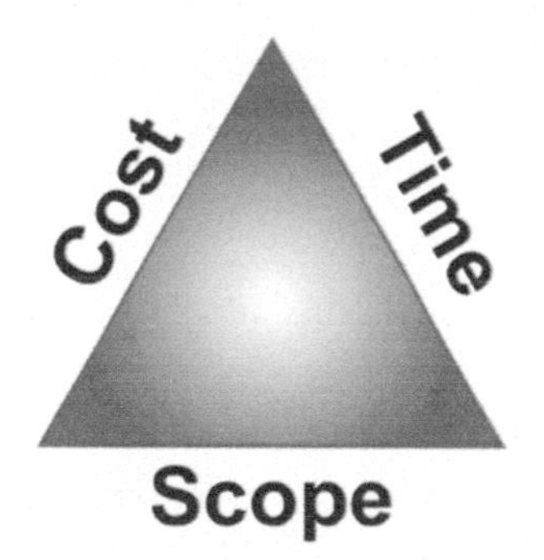

Figure 3-2.
Projects "Triple Constraint"

inadequate specifications, technical incompetence, poor communications, absence of support from upper management, changes and modifications—to name a few.

PROJECT MANAGEMENT PITFALLS

Projects fail for a multitude of reasons. Compromising any aspect of project delivery, i.e., time, cost, scope, can be a cause. And any concession on quality is simply unacceptable. To a great extent, they fail due to a lack of processes (planning, procurement, change, communication risks), to name a few. Without sound processes, project development becomes unpredictable and subject to breakdown of the project plan. A capital reason for project failure is a lack of clearly defined goals or objectives; they should exist at every phase and stage of the project lifecycle. Changing project requirements mid-stream can adversely affect project scope, schedules, and budgets. Inadequate resources, work overload and inexperienced project managers can also annihilate success, as can an inadequate budget or an unrealistic time-line.

A study performed by the United States Government Accountability Office (GAO) found that nearly 50% of the federally funded IT projects are either poorly planned or perform poorly on both of these. This led the government to pass new legislation to help shore up the government's project management efforts. The new law (S.1550—114th Congress (2015-2016): Program Management Improvement Accountability Act | Congress.gov | Library of Congress) was sponsored by Sen. Ernst, Joni [R-IA] (Introduced 06/10/2015) and on 12/14/2016 became Public Law No: 114-264.

The "Program Management Improvement Accountability Act" establishes as additional functions of the Deputy Director for Management of the Office of Management and Budget (OMB) to:

- adopt and oversee implementation of government-wide standards, policies, and guidelines for program and project management for executive agencies;

- chair the Program Management Policy Council (established by this Act);
- establish standards and policies for executive agencies consistent with widely accepted standards for program and project management planning and delivery;
- engage with the private sector to identify best practices in program and project management that would improve federal program and project management;
- conduct portfolio reviews to address programs identified as high risk by the Government Accountability Office (GAO);
- conduct portfolio reviews of agency programs at least annually to assess the quality and effectiveness of program management; and
- establish a five-year strategic plan for program and project management.

The Office of Personnel Management must issue regulations that: (1) identify key skills and competencies needed for an agency program and project manager, (2) establish a new job series or update and improve an existing job series for program and project management within an agency, and (3) establish a new career path for program and project managers. The GAO must issue a report within three years of enactment, in conjunction with its high-risk list, examining the effectiveness of the following (as required or established under this Act) on improving federal program and project management:

- the standards, policies, and guidelines for program and project management
- the strategic plan
- Program Management Improvement Officers and the Program Management Policy Council.

Other pitfalls include lack of management support (I'm glad to see the government finally taking action!), poor communications, and skill set mismatches. These can all be avoided through proper planning and incorporation of appropriate project management tools and techniques.

THE INFORMATION SUPER-HIGHWAY

Back in the day, before we had access to the internet, and organized project management was still in its infancy, there wasn't much information available on the subject; nor was it easily accessible. That all changed in 1991 with the advent of the "World Wide Web," that went live on August 6th of that year. At the time, most people (outside of the military) didn't even know what the internet was. Though the internet had been around for a number of years (with its own function and purpose), the "web" provided a means of accessing data online in the form of websites and hyperlinks. It has popularized the internet among the public, and served to develop the vast trove of information that we can now access on a daily basis. Now there really is a whole world out there that devotes itself to the understanding and advancement of the project management profession. Some sites require membership; some don't. Some of them charge a fee, and some are free.

There are a multitude of organizations, groups and individuals so dedicated. Sites can be found for certifying schools (like Villanova's Masters in Applied Project Management), scores of courses for certificate exam preparation (like Cheetah Learning's PMP exam prep course), case studies (like at PM Solutions) and industry reports (like the Standish Group 2015 Chaos Report). But, if I were limited to one particular organization from which to get all of my materials, information and instruction, it would be the Project Management Institute (PMI), for all the reasons cited earlier in this text.

According to Wikipedia, PMI provides services including the development of standards, research, education, publication, networking opportunities in local chapters, hosting conferences and training seminars, and providing accreditation in project management. PMI has recruited volunteers to create industry standards, such as "A Guide to the Project Management Body of Knowledge (PMBOK Guide)—Fifth Edition" (2013). Recognized by the American National Standards Institute (ANSI) as American National Standard BSR/PMI 99-001-2013. This will soon be replaced by the sixth edition in the fall of 2017. Other standards include the following.

Foundational Standards

- The Standard for Program Management—Third Edition (2013). Recognized by ANSI as American National Standard BSR/PMI 08-002-2013.
- The Standard for Portfolio Management—Third Edition (2013). Recognized by ANSI as American National Standard BSR/PMI 08-003-2013.
- Organizational Project Management Maturity Model (OPM3) | Knowledge Foundation—Second Edition (2008). Recognized by ANSI as ANSI/PMI 08-004-2008.

Practice Standards and Frameworks

- Practice Standard for Project Risk Management (2009)
- Practice Standard for Earned Value Management—Second Edition (2011)
- Practice Standard for Project Configuration Management (2007)
- Practice Standard for Work Breakdown Structures—Second Edition (2006)
- Practice Standard for Scheduling—Second Edition (2011)
- Practice Standard for Project Estimating (2010)
- Project Manager Competency Development Framework—Second Edition (2007)

PMI Standards Extensions

- Construction Extension to the PMBOK Guide—Third Edition (2016)

- Government Extension to the PMBOK Guide—Third Edition (2006)
- Software Extension to the PMBOK Guide—Fifth Edition (2013)

PMI's first credential was the PMP. It has since become a de facto standard certification. In 2007 it earned the ANSI/ISO/IEC 17024 accreditation from the International Organization for Standardization (ISO). As of 2016 over 710,000 people held the PMP credential. PMI later introduced many other credentials and a certification; including:

- Certified Associate in Project Management (CAPM)
- Project Management Professional (PMP)
- Program Management Professional (PgMP)
- Portfolio Management Professional (PfMP)
- PMI Agile Certified Practitioner (PMI-ACP)
- PMI Risk Management Professional (PMI-RMP)
- PMI Scheduling Professional (PMI-SP)
- PMI Professional in Business Analysis (PMI-PBA)

PROJECT MANAGEMENT TERMINOLOGY

The fact is there are literally enough definitions (depending on the depth of each explanation) of project terms to fill a book of their own. A more abridged version might form a healthy lexicon. But we're not even going after a glossary at this point. At this juncture, a few essential terms are all that are necessary to get you on your way. We'll cover the more arduous terminology as we proceed through the later chapters. We know what projects, project managers and project management are. That said:

Action Plan—A plan that describes what and when things need to be completed.

Activity—The work or effort needed to achieve a result.

Assumptions—Factors that are considered to be true, real, or certain.

Budget—An estimate of funds planned to cover a project.

Change in Scope—A change in objectives, work plan, or schedule.

Change Request—Requests to expand or reduce scope, modify costs or revise schedules.

Constraint—The state, quality, or sense of being restricted to a given course of action.

Deliverable—Any measurable, outcome, result, or item that must be produced.

Feasibility Study—A document that analyzes the technical feasibility of a project.

Gantt Chart—A graphic display of coordinated schedule-related information.

Identify risks—Determining which risk characteristics may affect the project.

Initiation—Committing the organization to begin a project phase.

Lessons Learned—The learning gained from the process of performing the project.

Milestone—A significant event in the project; usually completion of a major deliverable.

Monitoring—The capture, analysis, and reporting of project performance.

Process—A set of actions and activities performed to achieve a set of results.

Product—General terms used to define the end result of a delivered project.

Project Charter—The project manager's authority to apply organizational resources.

Project Plan—A document used to guide both project execution and project control.

Project Schedule—The planned dates for performing activities and meeting milestones.

Resource—Something to be drawn upon for aid or to take care of a need.

Risk—An uncertain event or condition that has an effect on a project's objectives.

Risk Mitigation—Revising the project's triple constraints in order to reduce uncertainty.

Schedule—The planned dates for performing activities and for meeting deliverables.

Scope—The sum of the products and services to be provided as a project.

Scope Creep—The addition of new requirements to the original product specifications.

Sponsor—The authority that provides the financial resources, for the project.

Stakeholder—Individuals and organizations who affect or are affected by the project.

Task—Well defined components of project work. Work package.

Team Member—The individuals responsible for some aspect of the project's activities.

Work Package—A deliverable at the lowest level of the work breakdown structure.

Chapter 4

The Project Toolbox

No one has a "knack" for managing projects. Proficiency in the profession is dependent on many factors, not the least of which is the utilization of the proper tools and techniques (and the capacity to apply them). Without those, even the simplest of endeavors is destined to fail. Consequently, many fledgling project managers are unable to successfully complete their projects, due to some deficiency in those mechanisms. There is an endless array of contrivances and approaches (hundreds) available for planning and carrying out project work, and the list gets longer as the undertakings get wider in scope, project type and difficulty. Too, as the engagement becomes more complex, the more project know-how is required to complete it. Call them tools and techniques, mechanisms, approaches or schema, but they all boil down to one category—"skill sets."

COMPARING SOFT AND HARD SKILLS

Project management requires a multitude of skills, all of which are necessary for planning and executing successful projects. Those broadly grouped skills fall into two main categories—soft skills and hard skills. Both are extremely important, but a balance needs to be maintained between the two. Soft skills without hard skills can cause the team to become complacent and inefficient, while hard skills without soft skills fosters an autocratic atmosphere giving the team no compulsion for personal achievement.

Soft skills (sometimes referred to as people skills) are interpersonal skills possessing personal attributes that are indicative of a high level of emotional intelligence, involving areas such as communication, leadership, motivation, patience, persuasion, negotiation and teamwork. They are applicable to most job titles in most industries, are essentially intangible, difficult to measure and personally employed, without the aid of tools.

Hard skills (on the other hand) are specific, teachable abilities that can be defined and measured, such as in areas involving planning, budgeting, scheduling, procurement, risk management, project monitoring, contract management and project performance. They augment the technical aspects of the project manager's role by amplifying a person's technical knowledge and their ability to perform distinct tasks. They assist with the development of tangible deliverables, through the use of such tools as work breakdown structures, critical path diagrams and earned value reports. Hard skills often involve the use of templates, spreadsheets and scheduling software.

A general (rule-of-thumb) distinction can be made between the two, as follows: Soft skills are often described using verbs that reflect human action or interaction; i.e., influencing, leading, decision making, problem solving, motivating, negotiating or communicating, etc. On the other hand, hard skills are generally described using nouns such as schedules, budgets, diagrams, metrics, reports or dashboards.

BUSINESS PRODUCTIVITY SOFTWARE

Business productivity software is any application people use to produce, create or modify data or information. Relative to project management activity, productivity software provides computer applications dedicated to producing documents, spreadsheets, presentations, databases, reports, graphs, illustrations and charts. There are a great number of singular and multi-function (both free and cost based) productivity software products on the market that can generate those fabrications mentioned above. Some are bundled into suites which bundle a group of programs that are sold as a package to solve common problems. My suite of choice is the Microsoft Office suite. MS Office includes word processing (Word), spreadsheet (Excel), relational database (Access) and presentation (PowerPoint) programs which hook up nicely with Microsoft Project (my preferred project management software). Microsoft Office is the most popular office suite and represents a large portion of the company's revenue.

Depending on its complexity, project management software can be used for organizing, planning, scheduling, cost estimating and control, budget management, resource allocation, decision-making, quality management and documentation. There are hundreds of computer- and internet-based project management software solutions available (some free, some for a fee) that are appropriate for almost every type of business. I have been using newer and newer iterations of MS Project for years to successfully plan and complete my projects, but (depending on your need and focus, other excellent available tools include Primavera, Easy Project, Wrike, Smartsheet, Project Kickstart and Team Gantt (again, just to name a few).

A picture (it is said) is worth a thousand words. Significant time, effort and expense can often be saved by substituting a drawing for a written description of a desired outcome. In project work, a rendering of a detail or footprint doesn't always have to come from a full-blown (expensive) Auto Cad program to get your point across to a contractor, architect or engineer. Sketches and diagrams can be drawn freehand or by using an inexpensive "Etch-a-sketch" type tool to develop the logic of a design, for conveyance to the professionals you've hired to complete the work. Such renderings can be a quick way to develop and record an idea, depict a small part of something at a larger scale, display complex construction details (such as a floor-to-wall junction, window or lighting layout) or show how component parts fit together. Countless programs exist on the market covering a wide range of sophistication and pricing (from amateur to professional levels) that are widely used by design firms and everyday consumers. I used such a tool on one project, to convey our idea for turning a set of exam rooms into a clean and dirty emergency treatment suite for the performance of unscheduled procedures. (See Figure 4-1.)

A number of other sketches were done showing the dimensions of the reconfiguration, pinpointing power and connections, laying out the plumbing and medical gas runs, spotting the HVAC diffusers and determining the ceiling grid and lighting plan. All of these were reviewed by the emergency services department for input and approval before being turned over to the contractor to perform the work and the A&E (architectural and engineering) firm for redlining of the existing (as-built) drawings. **Note**: For those who are uninitiated to the phrase, red-line drawings are intermediate drawings that show corrections or changes to a previous drawing. The term red line comes from the handmade marks using a red pen to make changes. The red-lined drawings are then used to develop (permanent) record drawings.

FORMS AND TEMPLATES

Forms and templates can be invaluable additions to your project management toolbox, both providing an apparent way to learn about project management and a straightforward means for accessing and working through the project management processes. Thanks to the existence of the internet and the many people and organizations willing to share their talents, time and resources, there is an infinite amount of information that is readily available on the subject of project management and access (much of it free) to forms and templates, that can be utilized to make its performance quicker and easier. Following are two examples.

Keeping with my allegiance to PMI and the PMBOK, I'll start with *A Project Manager's Book of Forms*, by Cynthia Stackpole Snyder. The book is a compendium of ready-made forms for managing projects, in line with PMI standards, written to assist both new and experienced project managers in handling all aspects of a project. It is a valuable asset that can serve as a resource

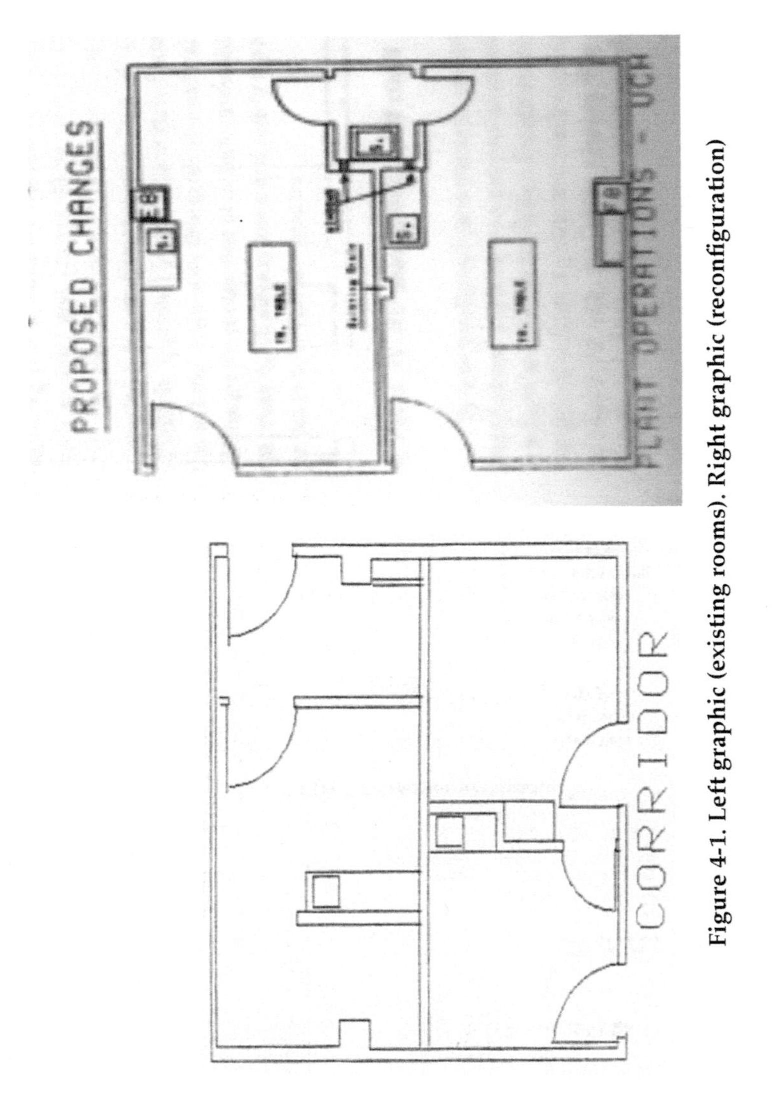

Figure 4-1. Left graphic (existing rooms). Right graphic (reconfiguration)

for collecting and managing project information. The forms are based on *A Guide to the Project Management Body of Knowledge (PMBOK® Guide)—Fifth Edition* and can be customized to meet the particular needs of a project or organization. PMI members have unlimited free access to the tools and templates found in this popular book, which can be accessed at projectmanagement.com where more than 1,000 templates are accessible to save both project time and effort. Of the 68 forms Snyder included in her book, I have found use for and value in the following titles:

Initiating Forms
- Initiating Process Group
- Project Charter
- Stakeholder Register
- Stakeholder Analysis Matrix

Planning Forms
- Requirements Documentation
- Project Scope Statement
- Assumption and Constraint Log
- Milestone List
- Cost Estimating Worksheet
- Responsibility Assignment Matrix
- Roles and Responsibilities

Executing Forms
- Team Member Status Report
- Change Request
- Change Log
- Decision Log
- Team Directory
- Issue Log

Monitoring and Control Forms
- Project Performance Report
- Variance Analysis
- Contractor Status Report

Closing Forms
- Procurement Audit
- Contract Close-Out
- Project or Phase Close-Out
- Lessons Learned

(Their URL is: http://www.projectmanagementdocs.com/#ixzz4m44hXPz7)

Author's note: For those of you who are not yet members of the Project Management Institute, there are a vast number of free on-line resources available to the general public. One of the better ones that houses a large (excellent) selection of free project management templates is a website titled PROJECT MANAGEMENT DOCS that is easily accessed through Google.

PROJECT MANAGEMENT PLANS

At some point in your life you've probably heard the old adage "If you fail to plan, you plan to fail." Nowhere does this state-

Figure 4-2. Power-point presentation cover sheet

ment hold more truth than when it is applied to the management of projects. Every project has a project management plan that describes how the project team will execute, monitor, control and close the project. Every project management plan is composed of a number of subsidiary management plans that when combined comprise and support it. It can receive information from all the subsidiary management plans and baselines. Here are the plan sections and their functions:

Change Management Plan

Most projects undergo changes from their originally intended configuration. This section describes the process used to control those changes. The impact of changes to any project must be carefully considered before they can be approved and enacted. Change control boards (CCBs) are often established by organizations to review and determine the fate of proposed changes. The change management plan identifies who can submit change requests, how they are monitored and who has the authority to approve or deny them.

Process Example:

- Any stakeholder can identify the need for a change by submitting a change request form to the project manager.
- The project manager logs the request into the change request register.
- An evaluation of the change request will be conducted by the project manager, project team and requestor to determine the impact of the change to cost, risk, schedule and scope.
- The project manager will submit the change request and analysis to the change control board (CCB) for review.
- The change control board (CCB) will approve or deny the request.

- If approved, the project manager will implement the change.

Communications Management Plan

Without adequate and appropriate communications with and between stakeholders, projects are doomed for failure. Communication management plans define the projects, the disseminating of information and how that information will be distributed. Project managers are responsible for ensuring effective communications on their projects, and for creating a plan for interaction between the project principals. The plan will serve as their guide throughout the life of the project and will need to be updated as communication requirements change. Thought should be given to who receives what information when (in a form that is clear, concise and timely), and a team directory should be created to provide the contact information of everyone involved in the project. Rules should be composed for when and how meetings are to be conducted, who will be required to attend them and how e-mails should be used and distributed. They should spell out the procedures for expressing any issues or concerns that arise. A deeper coverage of this subject will be forthcoming in Chapter 6.

Configuration Management Plan

According to the Project Management Institute, project configuration management (PCM) is the collective body of processes, activities, tools and methods project practitioners can use to manage items during the project life cycle. PCM addresses the composition of a project, the documentation defining it and other data supporting it. PMI has created a 61-page standard (a complimentary PDF download of this standard is available to PMI members). Each practice standard provides guidelines on the mechanics (e.g., nuts and bolts, basics, fundamentals, step-by-step usage guide, how it operates, how to do it) of some significant process (input, tools, technique or output) that is relevant to a project manager. Configuration management is key to a successful project.

Cost Management Plan

To ensure that you have an approach and methodology to managing costs throughout the life of your project, the cost management plan is used to define how the costs of a project will be managed throughout its lifecycle. Working within the cost management guidelines, it determines how project costs are measured, monitored, controlled and reported, to ensure that the financial aspects of the project are adhered to. The cost management plan identifies who is responsible for managing costs and has the authority to approve changes to the budget, disclose how cost performance is measured and reported, what formats the reports take, their frequency and who receives them. Again, it is the project manager's responsibility to manage and report on the project's cost, performance and deviations throughout the course of the project.

Human Resources Management Plan

Project managers manage projects, but they lead people. Proper management of human resources is a key to project success. The purpose of the human resources management plan is to ensure that appropriate human resources are acquired with the necessary skills, that resources are properly trained, that team building strategies are clearly defined, and that personnel activities are effectively managed. The human resource plan defines things such as roles and responsibilities, organizational charts, reporting relationships, how resources will be acquired, and when they will be needed. The plan may include an organizational chart, staffing plan, skill set requirements, training intent, how performance reviews will be conducted and the recognition and rewards system to be used. The project manager will review each team member's assigned work activities at the onset of the project and communicate all expectations of work to be performed. He or she will evaluate each team member throughout the project to record their performance and analyze how effectively they are completing their assigned work.

Process Improvement Plan

The purpose of a process improvement plan is simply to analyze a project's processes with the intent of improving them. Several steps are involved in creating a process improvement plan. Following them improves performance by expanding decision-making, enhancing the likelihood of achieving long-term results. Processes don't change by themselves; they break down when people deviate from the plan. Some of the most common causes of those breakdowns are inconsistent leadership and broken communication feedback loops. To establish a workable process improvement plan:

- Identify the core objective of the desired process improvement.
- Assign personnel who understand the time, costs, materials, and reporting constraints that will be required.
- Delineate your current process; describe how you want the modified process to differ from it, and establish a baseline for comparison.
- Develop a plan based on an analysis of the inability of the current process to meet its planned objective.
- Implement the new process and test its stability.
- Determine if the changed process has met your improvement criteria and either accept or reject it.

Procurement Management Plan

Only a well thought out procurement process will keep you from wasting project money (and time). An effective procurement management plan should clearly identify the required steps for procurement, from the project's beginning to its conclusion. It is the project manager's responsibility to ensure that the plan contributes to the successful completion of the project. When procurements are required, the project manager will work with the project team to identify all items or services needing to be procured. Within the limits of their authorized approval amount, project managers provide oversight and management for all project procurement, working with their project team, purchasing department and others. Over and above their authorized limit, procurements will likely need to be approved by the project sponsor.

It is the contracts and purchasing department's responsibility to review all procurement requests and determine whether to make or buy the resources, acquire the required services internally or externally, and begin the vendor selection process. If outside resources are indicated, the project manager will manage any purchased goods or services, measure the performance of any selected vendor and pass on any relevant information to the purchasing and contracts people.

Quality Management Plan

Initiating a plan to practice good quality management throughout a project will ensure delivery of an exceptional quality product. The management of quality in a project ensures that its deliverables conform to established standards of acceptance. A project completed without a quality management plan, may be considered substandard or unacceptable. To provide a foundation for monitoring and control, all project deliverables should be defined, required tasks understood and the prescribed work properly planned. The plan must depict and explain quality roles and responsibilities, quality control, quality assurance, and quality monitoring.

The project sponsor is responsible for approving all quality standards for the project. The project manager is responsible for implementing the quality management plan and managing project quality by ensuring all tasks, processes, and documentation comply with the plan. It is imperative that all variances of quality control and assurance be communicated to the project team and stakeholders. If any changes are proposed and approved by the project sponsor and CCB (change control board), the project manager is responsible for communicating the changes to the project team and updating the plan.

Requirements Management Plan

Project requirements are those conditions, capabilities or specifications applicable to a product or service that are required to satisfy a contractual obligation in a project. Managing those requirements is elemental to a project's success. Requirements management plans promote consistent analysis, documentation and management of all project requirements. As with most planning documents, the requirements management plan is a subsidiary plan of the project management plan and is created by the project manager. As all projects are different, it must be tailored to fit the needs of each project. The plan lists the various requirement categories and describes how they are planned, prioritized, tracked and reported. During their life-cycle, all projects will require some changes; these changes in requirements need to be managed properly. The requirements management plan provides for planning, analyzing, categorizing, prioritizing, quantifying, tracking and reporting of project requirements. The inputs (based on the processes set forth in PMBOK® Guide, 5th Edition) to developing the requirements management plan include the project management plan, the project charter, enterprise environmental factors and organizational process assets. (More information on inputs will be forthcoming in the Chapter 10 coverage of the PMBOK).

Risk Management Plan

There is risk in everything we do in life; how we mitigate it (or if we mitigate it) is of significance in project management work. Through the use of risk management, organizations can minimize the negative impacts of threats to its projects and maximize the upside impact of opportunities. Risk management is a professional field unto itself and prescribes a number of approaches for dealing with the risk reality. According to Wikipedia, risk management is the identification, assessment, and prioritization of risks followed by coordinated and economical application of resources to minimize, monitor, and control the probability and/or impact of unfortunate events, or to maximize the realization of opportunities. Risk management's objective is to assure uncertainty does not deflect the endeavor from the business goals (in our case the successful completion of our project).

Projects can fail at any point in their life-cycle phases, be it in design, development or execution. Several risk management standards have been developed to address risks associated with project work including the Project Management Institute's "Practice Standard for Project Risk Management" (this 128-page standard is available free to PMI members in PDF form). The practice involves planning risk management, identifying and prioritizing risks before they occur, conducting a quantitative risk analysis to estimate overall project risk, responding to identified high-priority risks, and monitoring and controlling for those risks and responses to them. They can include methodical processes for identifying, scoring, ranking and managing the various risks.

Every effort should be made to identify risks ahead of time, to implement a strategy for the management of those threats from the on-set of the project. Mitigation strategies include avoiding the threat, reducing the negative effect or probability of the threat, transferring all or part of the threat to another party, and retaining some or all of the potential for a particular threat.

According to the standard ISO 31000 "Risk management—Principles and Guidelines on Implementation," the process of risk management consists of several steps as follows:

- identification of risk in a selected domain of interest
- planning the remainder of the process
- mapping out the following: the social scope of risk management, the identity and objectives of stakeholders, the basis upon which risks will be evaluated, constraints, defining a framework for the activity and an agenda for identification, developing an analysis of risks involved in the process, and mitigation or solution of risks using available technological, human and organizational resources.

Schedule Management Plan

It is the project manager's responsibility to create a schedule management plan to ensure the effective management of a project's schedule throughout its life-cycle. A well-planned and properly developed schedule management plan is the roadmap to ensure that that goal is met. The development of a schedule can be likened to the development of a project in and of itself—it having a temporary nature, with a start and end date, definite goals and multiple stakeholders. Those stakeholders should have a clear understanding of the terminology used in the schedule and in the schedule reports. Technical scheduling terms such as critical path (CP) and work breakdown structure (WBS) should be defined so that phrases like the "longest path," "total float" and "work package" are recognized. A good understanding of all schedule-related terms will enhance stakeholder communications and promote better adherence to the project schedule. The schedule management plan should include a list of documents such as contractual information, plans and specifications, resources, penalties, incentives and how the schedule is to be used. It is also important to identify which stakeholders and end users have input into the project or the schedule. Once identified, their roles and responsibilities should be spelled out and clarified.

Author's note: The PMBOK® Guide notes the use of the RACI Chart, correlating tasks with roles and individuals by identifying four links: responsible, accountable, consult, and inform. This covers the major divisions among project management team members in the development and maintenance of the project schedule and will provide clarity to the project team.

Scope Management Plan

The scope management plan (SMP) defines, develops, details and verifies all of the tasks required to complete a given project and establishes who is responsible for managing and controlling the scope. It documents the scope's approach, definition, verification, control measures and work breakdown structure. Project scope management is the responsibility of the project manager, with the sponsor and project team playing key roles for establishing and approving documentation using quality checklists and work performance measurements. A project's scope is built through a five-step process wherein the project's objective requirements are defined, detailed product descriptions are developed, a work breakdown structure (WBS) of work packages is created, the scope is accepted by the sponsor, and the process is monitored.

The project charter is the basis from which the project requirements are defined and clarified, enabling their measurement once the project begins. The product description establishes the project framework and includes all of its assumptions, constraints and deliverables. The WBS simplifies the scheduling, costing, monitoring, and control of the project, breaking the deliverables into smaller (more manageable) work packages. In the acceptance phase, the scope is verified and baselined before being accepted. Continuous monitoring of the scope allows for control of any variance from the established baseline (scope creep). Scope changes may be initiated by the project manager or stakeholders or any member of the project team. Change requests are submitted to the project manager to provide estimates and impacts to project schedules and costs. After evaluation, the PM submits the scope change request to the change control board and project sponsor for acceptance. When properly implemented, an SMP helps manage the time, cost and quality. Well executed scope management ensures that all (and only) the required work necessary for completing the project is included in the project.

Stakeholder Management Plan

The stakeholder management plan is a subsidiary plan of the project management plan. As is the case with all project planning documents, the plan is created by the project manager. Its purpose is to identify the people, groups and organizations that could affect or be affected by the project, analyzing their expectations and impact on the project; and to develop appropriate strategies and tactics for effectively engaging stakeholders in a manner appropriate to their interest and involvement in the project. It defines the requirements, processes, and techniques for engaging stakeholders, based on an analysis of their needs, interests, and abilities to influence the project. It is also used for determining the types and amounts of communications each stakeholder should receive and includes strategies for managing both. The plan's objectives are to document and communicate how information will be disseminated to, and received from, all stakeholders. It identifies who the stakeholders are, the requirements of each, the means, frequency and duration of communication, and the roles and responsibilities of the project management team.

After identifying a project's stakeholders, all the information gathered should be entered into a stakeholder register. A stakeholder register is a project management document which contains all of the data gathered about the group. In the register you can find stakeholder names, titles, roles, interests, power, requirements, expectations, and type of influence, etc. Once the register is created, a strategy can easily be drafted to manage them. The register will contain identification, assessment and classification information about each stakeholder, as well as their individual requirements including:

- Communication needs
- Communication frequency
- Expectations
- Influence on the project
- Interest and power

The stakeholder register must be updated when any new stakeholder is identified or any change is observed in stakeholder attributes. As stakeholder identification is a continuous process, you will need to update the document throughout the project life cycle. As the register contains names, e-mails (and addresses), stakeholders' classifications, and the strategies to manage them, it should be kept in a secure place, and its access should be restricted.

ACTION LOGS AND REGISTERS

It goes without saying that, in project management, paperwork is an unavoidable and necessary evil; the more complex the project, the more information needs to be collected, reviewed and processed. Keeping it organized is the secret to dealing with it all. This is where the value of project logs and registers comes in. They are tools that can be used to break projects into manageable and coherent informational groupings to address critical tasks, target dates, responsibilities and resolutions. Too, they can be used as placeholders for tasks that need to be accomplished, as presentation documents for decision making, and to record completed work for the filing cabinet. Along with identifying which tasks need to be done, they spell out how much time each will require, determine what it will cost, and assign the parties responsible for completing them. They can also provide a forum for addressing project issues and determine what resources are needed and available to bring them to closure. Project logs and registers enable project managers to constantly review a project's status with their team (and other project stakeholders) and to avoid misunderstandings that would otherwise go unattended. Not all action registers are the same, however. If poorly executed, action registers can mean more tasks to be taken on unnecessarily by the project team. Instead of focusing on the main project tasks, tasks such as manual checking, or cross-checking and updating of data can drain the team's valuable energy and time. Here are four logs/registers I use frequently in my projects:

Assumptions Log

An assumption is something taken for granted—a supposition or a belief (based on your knowledge, experience or the information at hand) of what you assume to be true for the future. Assumptions are important considerations of any project. In project work, they are anticipated events or circumstances that are expected to happen during a project's life cycle, which will require validation and follow-up, to determine what impact (if any) they will have on that project. An assumption log is a document which enables capture, tracking and documentation of those assumptions, allowing teams to identify which may carry risk and ensuring that there is a clear justification for decisions made about them. The log should assign each assumption a reference number, describe each assumption, provide the assumption status, and prescribe any validation or follow-up actions. Each assumption should have an owner responsible for its follow-up and for validating it, and the assumption log should be updated as items are closed or more information is obtained.

Risk Register

All projects include uncertainty over what will happen during their life-cycle, and every assumption reduces the chances for a project's success. It is the responsibility of the project manager to ensure a project's success by managing the risks that are imposed on it. Risk registers are created early in the project and are used to record identified project risks, their status and history. They evolve over the project life-cycle as risks are reviewed, revised, and updated—as the project progresses. The registers enable project teams to take action on risks before they have an opportunity to become "issues" that can have a negative impact on the project. A best practice is to focus on the causes of the risks, determine their probable impacts and create a standardize process for dealing with them. A simple analysis of the impacts can help to classify risks and prioritize the actions needed for their mitigation. It is important to note that when deviations occur in the plan, risks become issues that must be addressed to keep the project on track.

Issue Log

Issues are basically any conditions that reduce the team's ability to execute the project plan. They are easily identified because they cause obvious schedule slippage and extra work. The issue log is where problems are recorded that were not accounted for in the original plan and that threaten to delay the project, push it off budget or affect its scope (triple constraint). Project managers use issue logs to capture and monitor information on a formally managed variety of problems, inconsistencies or conflicts including change requests, performance reductions, potential new risks, etc. A detailed log will provide clarity and a point of future reference for actions, comments and resolutions of issues.

The issues log records details of all the issues identified at the beginning and during the life of the project, the action taken to address each issue and the subsequent results. The information it contains includes issue descriptions with individual ID numbers, who raised the issue, the person or group responsible for its resolution, an open/close status, and how the issue was resolved.

Change Register

As previously stated in the change management plan, most projects undergo changes from their originally intended configuration. Change management is one area of project management which will cause serious problems if it is not carefully managed. Without a deliberate change management effort, changes in design, scope, deliverables, and various other changes can easily cause a project to fail. Failing to track changes and consider their impact before they can be approved and enacted can cause critical mistakes to be made that can result in poor project performance. The register is a document wherein all notable changes made to a project are recorded. Logging changes into a register helps to keep processes organized and projects on track.

The change register is used by the project team to log and track change requests throughout the life of the project. It should be updated as new change requests are submitted or as existing change requests are approved or rejected. The information placed in the register will come directly from the change request form. The register should be updated as needed and all changes clearly communicated to all stakeholders.

Submittal Log

Submittals refer to the written or physical information provided by a responsible contractor of documents (such as schedules) and/or physical items (such as ceiling tile samples) that contractually require the approval of stakeholders. Submittals in construction management include matters such as shop drawings and blueprints, material data and samples, cut sheets on equipment, product manuals and test reports. The project manager is responsible for creating the initial submittal schedule which contains all of the required project submissions (including specifications and due dates). As submittals are received from the contractors or vendors, they are added to a submittal package for review. That information is given to the design team for review and approval of equipment, materials, etc., to ensure that the information complies with the project drawings and specifications before they are fabricated and delivered to the project. Once approved, the submittal is returned to the contractor, signifying that the work (or fabrication) is approved for construction. The submittal log will be where all specification requirements will be cataloged, dates of performance identified and actions by responsible parties summarized.

OTHER TOOLS AND TECHNIQUES

The above may seem like a comprehensive addressing of the subject, but it is far from all inclusive. Most of what was covered deals primarily with the "hardcopy" paperwork side of the project toolbox. There are hundreds more tools and techniques available for use, and they are limited only by the particulars of the project you will be managing. There are also mathematically based tools such as the critical path method (CPM), project evaluation and review technique (PERT), work breakdown structure (WBS), and project formulas for determining such things as earned value (EV), planned value (PV), cost variance (CV), cost performance index (CPI), estimated at completion (EAC), schedule performance index (SPI), variance at completion (VAC), estimate to complete (ETC), to complete performance index (TCPI), and others. As we are in the novice section of the book, this subject matter is simply too heady a read at this juncture. It will be covered more thoroughly in later chapters as we advance our thought process into the more appropriate practitioner level.

Part II — The Coordinator

Chapter 5

Assembling the Project Team

PROJECT MANAGERS ARE LIKE SHIPS' CAPTAINS whose crew members are the project's management team. Now there are good captains, like John Paul Jones ("I have not yet begun to fight!"), and bad ones, like William Bligh (of *Mutiny on the Bounty* fame). The men they commanded were just as different from one another as their stories. Jones' crew was comprised of able-bodied, experienced seamen and disciplined marines; while Bligh's hands were a rag-tag group of inexperienced sailors, unused to the rigors of the sea and corrupted by the loathness they acquired during their 5-month layover in Tahiti. Unfortunately, neither Jones nor Bligh picked the people who reported to them, both being assigned to

Figure 5-1. They languished 5 months in Tahiti, waiting for the breadfruit trees to ripen!

ships having a full complement of men already aboard when they took command.

Project managers, on the other hand, have a better chance of navigating their journeys than those two, because they have the latitude to choose their team members before venturing off into uncharted waters. Bringing together a collaborative, cooperative and committed group of competent self-starters can make for smooth sailing; while choosing unwilling, incompetent, conspiring and self-indulgent individuals might well sink the ship.

PROJECT TEAM ROLES AND RULES

It's not always apparent who a project's team members are, or what roles they play, but it's certain that the project manager is the primary player and responsible for the successful completion of the project. While managing relationships with and within the project team, given sufficient resources, the PM is responsible for meeting project objectives and ensuring that it's completed on time and within the budget. Among their many duties they are also responsible for assembling, leading, assigning tasks to and managing the project's team members. A project team may be comprised of in-house staff members or external consultants who work on one or more phases of the project on a full-time or part-time basis. Their duties may include providing contributions as subject matter experts (SMEs), conferring with and establishing the needs of the end users, documenting processes and completing individual deliverables. Ancillary team members are the project sponsors, who approve the budget, ensure the availability of adequate resources and make key business decisions for the project; and business analysts who are responsible for documenting the project's technical and business requirements and verifying that the deliverables meet the project's objectives and requirements. Both of them work closely with the project manager.

Team building is all about selecting the right individuals for the job, molding them into a cohesive group and building relationships between them. The project manager needs to understand the strengths (and weaknesses) of each potential team member and what value they bring to, or what detriment they might have on, the project. Once selected, the PM needs to put rules into place to ensure that everyone will know what is expected of them and how information is to be communicated. To be most effective, all team members should participate in the team-building and goal-setting processes. The number of members on a team, their level of expertise, and the time they have available to devote to the project can depend upon several factors; i.e., the goals of the project, timing and cost considerations, whether they are staff members or contractors, who they currently report to, etc. Here are some scenarios I've worked with on past projects:

- Remember my first? The majority of the team was comprised of members of my maintenance staff. But I couldn't have completed the parking lot without input from the respiratory therapy department (re: the medical gas cylinders), the receiving department (re: re-routing of product deliveries) and representatives of the affected vendors and contractors.
- In Chapter 4 I referred to an emergency treatment suite that my department designed. Again my staff had some involvement, but as all of the construction work was to be performed by a general contractor (under the auspices of an architectural firm), I only needed one (master) mechanic on the project team (to assess what impact the proposed changes would have on the existing systems). In that case, the emergency services department was represented (by a physician) and the architect and GC were members of my team.
- As I took on more complex projects throughout my tenure as PM, the composition of my project teams changed accordingly; ergo, when I conducted a business office department relocation, I included some staff (for electrical and plumbing work), a coordinator from the business office, and representatives from the telephone service and moving companies.

- Later (more sophisticated) projects called for the inclusion of estimators, inspectors, schedulers, office personnel, regulatory body representatives, specialty vendors, sub-contractors, quality assurance managers, accountants, administrative personnel, purchasing agents and many others.

BUILDING AN EFFECTIVE PROJECT TEAM

What does it take to build a strong and effective project management team? My guess is a strong and effective project manager! The ability to form an effective team is a key competency of a successful project manager. In my book, a competent team builder is someone who has been down that road before and has learned from their experiences, either as a successful project team member or as a fledgling project manager who vowed to do better during their next go round. You need not have experienced the "thrill of victory" nor suffered the "agony of defeat," but what you will need is to have a good grasp of what it takes to coach a group of individual thinkers into exerting a unified effort. When selecting your team, it is important to choose the right people. Following is my roadmap for selecting the appropriate individuals and building a successful team.

Qualifications and Abilities

Member qualifications and abilities don't necessarily need to be stellar, but their skills sets should absolutely reflect the needs of your project. In other words, don't hire a master electrician if all you're doing is installing ceiling fans. And be certain to draft a team with a variety of skill sets rather than folks having identical know-how, such as individual trades personnel (plumber, carpenter, and electrician, etc.) as opposed to a group of general maintenance mechanics. The diversity of their knowledge base may well contribute to a faster completion time or better use of resources. At the same time ensure that, as a group, all of the skills that they bring to the table are proportional to those that are needed for the project (in both level and amount). Choose people who can get along with each other and work together as a team.

Goals and Ground Rules

At the start of your project, clarify who is in charge and what you expect from your team. Establish a meeting schedule, and enforce attendance and recording of minutes (late is bad and tardy is worse). Ensure that everyone understands the team hierarchies and reporting structures and what the rules are for dealing with problems and conflicts. Set unambiguous, clear and achievable goals for both individuals and the team. Make certain the goals are easily attainable and designed to build team spirit and enthusiasm and to establish a sense of pride in their achievement. That is best accomplished early on in your project while your team is still getting organized. Experiencing some early successes will bolster their morale and provide them with the confidence to take on tougher assignments as the project progresses.

Communications

Nothing can sabotage a project's successful undertaking more than poor or inadequate communications. If the team isn't totally aware of the entire goings on, they will have no way of dealing with project conditions or situations and will be unable to affect its status. It is the PM's responsibility to ensure that the team is apprised of and kept up to date on all project activities, team interactions, any potential conflicts and all approved changes. Problems or delays in one area can cause havoc in others. To avoid them, set a tone of cooperation and discourage competition. The subject of effective communications will be detailed more thoroughly in Chapter 6.

You've probably surmised by now that team building is not a singular endeavor, but rather an ongoing effort by the project manager to constantly refashion the team throughout the lifecycle of their project, which is crucial if the project is to succeed. The PM needs to bring a complete set of soft skills to the job—knowing what intensity needs to be applied for a given situation,

Figure 5-2. One for all, and all for one!

whose hands need to be held, how to acknowledge achievements and (if necessary) when to "crack a whip." Comradery among team members begets collaboration, which begets better communication, which culminates in a job well done. The way that project managers interact with their teams is of paramount importance and can have a bearing on their ability and willingness to contribute.

THE FIVE TEAM DEVELOPMENT STAGES

So, how do those folks get together and in sync? Fortunately, others had been confronted with that scenario in the past and worked out some of the kinks, making it easier for the rest of us to grasp. Thanks to Bruce Wayne Tuckman (professor of educational psychology at Ohio State University) and Mary Ann Jensen (who later collaborated with him to expand his theory), we know that every project team goes through five stages of development. The first four stages of the team development model (Stage 1: Forming; Stage 2: Storming; Stage 3: Norming; Stage 4: Performing) were proposed and developed by Tuckman in 1965. In 1977 (with Jensen's help) that model was revised to add a fifth stage, Stage 5: Adjourning. The adjourning stage occurs when the team is completing the current project.

Regardless of size, every team goes through the stages of development. Whether the projects are simple or complex, all must progress through the five stages, without exception. To quickly ensure the formation and management of an effective team, for meeting the project's objectives, the project manager must ease the team through each of the stages. The PM must guide their team effectively, providing strong guidance and direction early on, while allowing the team to solve their own problems and resolve any differences they may have, as they move through Tuckman's stages. As mentioned, a team can be comprised simply of staff members or pulled together from various parts of a large organization and can have any number of added members from outside sources. Once chosen, the "team" will go through the stages, as follows:

In the *forming* stage, team members first meet each other and share information about their backgrounds, interests and experience. They learn about the project, discuss its goals and objectives, probe into where they fit into the project, and concern themselves with how their capabilities and skills compare with one another. Feeling each other out, they will test each other and look to the project manager for direction. Here the PM clarifies the team and project goals and works with the team to establish norms for working together.

In the *storming* stage, the team begins to work together, and its members jockey for position, establishing their individual power bases to determine how they will work together. The team members will compete with each other for status and for acceptance of their ideas. Here they set down the rules that they will follow to work together and resolve their differences. The team members also start to make significant progress on the project as they begin working together more effectively. The PM's role is to foster good communications between the members and ensure that they maintain the project's focus. At this point, the PM should allow the team more independence and start depending on them for some decision making.

In the *norming* stage the team will begin to work more effectively as a team, becoming conscious of their existence as a cohesive group and accepting the rules that they have established for themselves. They respect each other's opinions, see the value of their differences and consistently work toward a common goal. At this juncture, they become more trusting of one another and work together more effectively. The team is highly motivated to get the job done. They can make decisions and problem solve quickly and effectively. The project manager ensures that the team promptly resolves any internal issues and maintains their single-minded purpose. The PM can become less engaged in the team's day-to-day decision making and problem solving as the team becomes more seasoned.

In the *performing* stage, teams function at a very high level. The focus is on reaching the goal as a group. Team members work effectively as a group and do not need the oversight that is required at the other stages. The members have gotten to know each other, trust each other and rely on each other. The individual members of the team act as one. Everyone is on the same page and focused on reaching the project's goals. Each depends on the others to deliver the goods uniformly. They are a team in every sense of the word and have less need for intervention by the project manager. The PM's job now is to act as a go-between and keep the stakeholders abreast of the team's performance.

In the *adjourning* stage, the team is dismembered and will be joining other teams and moving on to other work. The members have a feeling of loss about having to move on, going their separate ways, and they will miss the interactions they shared during their group experience. The project manager should have the members engage in some sort of activity to bring closure to the project and ease the transition.

TEAM DEVELOPMENT TOOLS AND TECHNIQUES

According to the PMBOK, the tools and techniques needed for the development of a project team are: interpersonal skills, training, team-building activities, ground rules, collocation, recognition and rewards and personnel assessment.

- *Interpersonal skills* (soft skills) are behavioral competencies that enable the PM to communicate, express emotions, resolve conflicts, influence people, and build teams. They enhance a project manager's capacity to empathize with people, as well as boost their ability to negotiate with vendors and communicate with stakeholders. They are different than general management skills that include such abilities as general business knowledge, leadership, budgeting acumen, organizational comprehension and problem-solving aptitude.

- *Training* is called for when project team members lack the necessary management or technical skills needed to complete their duties. Those skills can be developed as part of the project work and can be formal or informal in nature. Methods may include computer-based, classroom or on-line instruction, on-the-job training or mentoring by a subject-matter expert (SME). In all cases, training includes all activities designed to enhance the competencies of the project's team. Member's skills and abilities are assessed against project needs, and training is provided that supports the members' ability to carry out their assigned tasks. If it's known early in the planning processes that training will be necessary, the need should be detailed in the staffing management plan. The plan should be updated when training requests are granted or when training is called for as a consequence of a performance appraisal finding.

- *Team-building activities* are intended to help individual team members work together effectively. As the project manager, it's your job to make certain that the team members understand the project goals and their individual assignments. T-B activities are also a means of getting team members to know, understand and accept the goals and objectives of the project and an opportunity for the PM to address and clarify any misunderstandings the members may have regarding the direction of the project. Activities can occur in a variety of ways and settings, from a short discussion in a status review meeting to a formal program designed to improve the interpersonal relationships between the team's members. Such activities can help in building essential trust and comradery that is so important to establish during the initial stages of a project. Project managers need to continually monitor their team's functionality and performance, then create and conduct activities to prevent or correct any found problems.

- *Ground rules* are precepts or instructions, set by the project manager and project team that describe acceptable team behavior. Their formulation gives the team clear guidance on what is expected from them during their interactions, decreases misunderstandings and improves team productivity. The rules cover such concerns as work interaction, code of conduct, ethics and etiquette, conflict resolution, requests for training, meeting attendance, report generation and communication. Once the rules are established and agreed to, all project team members share responsibility for their compliance and enforcement.

- *Colocation* enhances the ability of project team members to perform as a team by placing the most active (if not all) of the members in the same physical location. Colocation can be temporary (such as at strategically important times) or permanent (for the duration of the project); depending on project needs, budgetary constraints and available resources. And it enables teams to function more effectively than if they're disconnected geographically. Often, project managers house the team in a large room (sometimes referred to as the "war room"), designed and reserved specifically for the project team to meet and work in. While colocation has its advantages, using virtual teams (via the internet) can benefit a project through lowered costs, less travel, and a closer proximity of team members to suppliers and key stakeholders.

- *Recognition and rewards* is integral to motivating people to work more efficiently and produce better results. Team members should be recognized for exemplary effort and rewarded for extra performance, proportional to the achievement. Motivation can be extrinsic (material rewards) to include cash bonuses, stock options, the use of a company car, etc., or intrinsic (a sense of accomplishment for a job well done). People need to feel that they are valued in the organization and that their efforts are appreciated. Most team members are motivated by an opportunity to apply their professional skills to meet new challenges. The PM is tasked with developing the criteria for rewards, especially monetary awards. Considered a positive aspect of managing a team, rewards can have a negative impact on the team if a proper method isn't established for handing them out. It is important for the project manager to recognize the team's contributions throughout the course of the project's life cycle and not wait until the closure phase to thank them for a job well done.

- *Personnel assessment tools,* such as surveys, assessments, interviews, ability tests, and focus groups help project managers gain insight into a team's areas strengths and weaknesses. They enable the assessment of a team's attributes and how they process and organize information. PA tools can provide improved understanding, trust and commitment among team members and cause them to be more productive. Reasonable and accurate application of the tools can result in improved competencies, enhanced project performance, reduced staff turnover rates and motivated team members. Again, it is the project manager's responsibility to create, choose and/or install and utilize the tools for the benefit of their teams and success of their projects.

PROJECT TEAM CHARACTERISTICS

Effective teams have a clear unity of purpose and are typically very energetic and cohesive groups that form their own particular set of characteristics. Their enthusiasm is contagious, as evidenced by their inherent productivity and problem solving abilities. When presented with a goal, they work diligently as a single unit to complete it in a timely and cost effective manner. Each individual on the team is self-conscious about its operations and carries his or her own weight. In an informal, comfortable, relaxed atmosphere, they share a clear, concise and mutually agreed-upon approach to handling their objectives. To create a productive team, you must first be able to identify the characteristics of effective teamwork. Characteristics common to such teams include:

- common values and beliefs
- support for one another
- commitment to the project
- agreed-upon processes
- a focus on end results
- support for the organization
- enhanced communication
- objective evaluation of ideas

- a trusting, open attitude
- the ability to resolve conflicts
- information and sharing
- measurable objectives
- a self-evaluation process
- understanding of expectations
- continuous monitoring of performance
- a sense of belonging and purpose

As the team's leaders, project managers know that such characteristics will not manifest themselves without their participation. It is important for them to espouse the values and beliefs they want the team to represent and exercise, while setting examples of them in their own actions. They should encourage their teams to openly share information, hold frequent meetings, and direct their energies toward achieving the mission's goals

Figure 5-3. Poor team characteristics result in team in-fighting

and objectives. PMs must engender high morale by establishing an esprit de corps within the team's ranks. They must encourage independent thinking, empowering team members to complete deliverables to keep the project moving forward. While instilling those virtues, the PM must be careful to avoid instituting bad ones. An ineffective team is unfocused, conflict ridden and filled with distrust. Such teams have low expectations, high turnover rates and lack the tools they need to achieve their goals (if indeed any have been set).

Some poor characteristics include:

- vague role definitions
- non-existent behavioral rules
- unrealistic expectations
- unresolved problems
- a lack of openness and trust
- insufficient resources
- frequent team infighting
- micro-management of efforts
- no feeling of commitment
- self-indulgent interests
- an unfair reward system
- unclear assignments
- poor reporting structures
- miscommunications
- ostracized team members
- no performance evaluations
- poor reporting structure
- a low tolerance for diversity
- lack of management support

Chapter 6

Effecting Project Communications

WELL TIMED, EFFECTIVE COMMUNICATIONS is an essential component of successful projects. It is evidenced when stakeholders speak the same language (project management has its own specific terminology to deal with), and when they appear to be on the same page (are kept informed and up to date). Whether communications is a tool in and of itself is up for debate, but it does utilize a great number of tools and techniques to educate the stakeholders, get their points across and keep them apprised of a project's status. Whole books have been written on the subject; I will do my best to give you a solid foundational understanding in this abridged chapter.

Much of what I present here will have a basis in the PMBOK (5th edition) Project Management Knowledge Area: Project Communications Management section. That section covers the planning, managing and controlling of project communications where the PMs take their guidance for writing the project's communication plan and monitoring all the incoming and outgoing stakeholder communications. A variety of communication methods (verbal, written and electronic) can be used to deliver project information to collect data, track issues, assign tasks, plan work, report status, document performance, distribute information, resolve conflicts and record data. Communications can be one-way (linear and limited because it occurs in a straight line from sender to receiver and serves to inform, persuade or command) or two-way (which includes feedback from the receiver to the sender).

The PMBOK® Guide identifies three types of communication; i.e., interactive (two-way), push communication (when the sender gives out information but is not looking for an immediate response), and pull communication (which provides stakeholders access to common project information). Directives, memos, letters, e-mails, faxes and reports are examples of push communication. Pull communication is evidenced by the use of information posted on or in repositories such as websites, bulletin boards and knowledge bases. Interactive communication allows for an exchange of ideas between active participants, who can have an effect on one another and is the project manager's preferred *mode* of communication. And "assertive" communication (whereby individuals clearly state their opinions and feelings without violating the rights of others), is their preferred *style*. The guide helps project managers ensure that the communications needs of stakeholders is understood and determine what communication outputs will be exchanged over the course of the project (i.e., status updates, minutes of meetings, reports on deliverables etc.).

CHANNELS OF COMMUNICATION

In a project, information flow is referred to as communication. Communication channels refer to the way this information flows within the project and between the project's principals. Breakdowns in project communications lead to an inefficient flow of information. It is the project manager's responsibility to implement effective communication channels to optimize team member productivity and to ensure the smooth running of their projects. Whether the information flows upwards, downwards or sideways depends on the position of the communicator in the communication web. Ergo, information from project team members may flow sideways to their peers or upwards to the project manager, whereas information from the project manager may flow down to their team members or up to the project sponsor. For the flow of information to be effectively communicated, there must be channels in place to ensure that it is properly and accurately transmitted and received. The channels enable the PM to provide stakeholders with comments and direction and are a feedback mechanism for two-way communication.

There are three main types of channels: *formal* (for transmitting and/or transferring information such as goals, directions, memos, reports, and meeting schedules); *informal* (discussions with other team members, the project manager and vendors or clients outside of a format setting); and *unofficial* (the project grapevine during work sessions and social gatherings) where both good and bad, true and untrue information is circulated. Informal communication is a part of every project but can cause problems if not monitored and controlled. Bear in mind that any issues, concerns, or updates that arise from informal discussions must be communicated to the project manager, so that appropriate action will be taken to deal with them. Aside from face-to-face encounters, hardcopy transmittals, phone calls, emails, faxes, logs, files and databases, other types of channels (medium) include; websites, video conferencing, project software, electronic bulletin boards, and mobile technology.

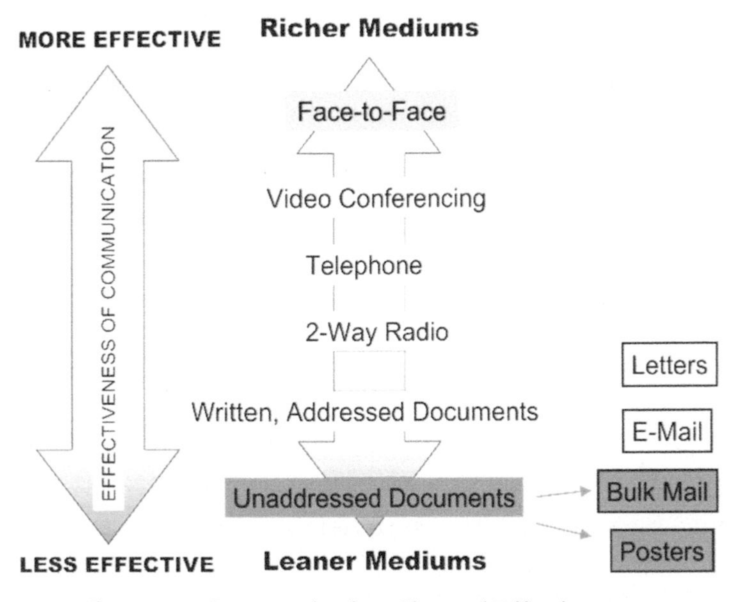

Figure 6-1. Communication Channel Effectiveness

CREATING A MANAGEMENT PLAN

Considerable thought should be given by project managers and their teams as to how communications will be managed throughout the course of a project. Communications plans are created during the planning phase, then executed and monitored over the course of a project's implementation. Instead of developing their own, many project managers opt to either embrace one that they've used on a prior project, follow a pre-ordained procedure instituted by their project management office (PMO), or conduct their communications on the fly. A solid communications management plan defines the project's informational requirements, spells out who is responsible for the flow of what information, when information transmittal or exchange must occur, and how all project information will be distributed, monitored and recorded. The more comprehensive the plan, the less problems you'll experience. To create a communications management plan, the audience and requirements must be defined, the medium for communicating must be chosen and a communications schedule must be built. The defined audience is a listing of the key stakeholders needing information about the project before and during its execution. The requirements establish what the key stakeholders need to know. To ensure that the stakeholders are kept apprised of the project's status and that the needed information is promptly and accurately delivered, a communications medium should be selected that is suitable for the information being conveyed. The communication schedule should be prepared and verified by the stakeholders who will be relying on it for direction as to when and where information exchange is to take place. The content of all communication should include the purpose of the communiqué and the responsibilities of the senders and recipients for dealing with the information being transmitted or received. In all cases it should be reviewed by the project manager before being presented to the stakeholders.

It is generally accepted that there are three categories of

communication that need to be tracked in a project communication plan; i.e., meetings, reports and records. In my view, *meetings* are the most effective way to distribute information. When planning a meeting, the project manager should carefully consider exactly what the objectives of their communication are and choose a meeting format that will best convey those to the attendees. Should the meeting be conducted face-to-face, electronically or via tele-conference? Determine which people need or want what information and at what frequency it should be given. If the meeting is with members of the project team, depending on the project and team size, it can be conducted on a conference call, as formal meetings, by way of email chats or simply through one-on-one conversations. Such exchanges keep the project manager to up to date on the project's status and allow for the resolution of issues before they get out of hand. If the meeting is with other stakeholders, it's always a good idea to get some face time so that you can take advantage of the captive audience and any clues that might be derived from witnessing the participant's body language. At that level it is best that you come to the meeting well prepared for the discussions that will ensue. From time to time, as the project moves forward, it is incumbent on the PM to provide periodic *reports* for the management team. Their purpose is to describe the project and its status to the persons chosen to review them and to provide historical documentation. As the recipients of those reports may not be familiar with the goals, risks or other particulars of the project, they need to be able to stand on their own and present the project in its entirety at the time of its issuance. *Records* of the project are used both during its undertaking and for subsequent scrutiny after the fact. Archived information in a project's files can be, and often is used to learn from the mistakes made while carrying it out, for planning and implementing new business strategies or as documentation needed to defend against litigious allegations relative to the fulfillment of a contract. Consequently, they need to be accurate and complete so that they convey a total picture of the actions that they represent.

TALKING IT OVER (MEETINGS)

Every project requires periodic meetings to be held, be they ad hoc conversations between project managers and their management teams, informal discussions between team members or full-blown progress review sessions with all stakeholders in attendance. Meetings of all kinds are held to discuss the project's goals, assigned tasks, problems and progress that directly contribute to or detract from a project's overall success. It's up to the project manager to determine the type and frequency and to create an agenda for each meeting. Meeting types include:

- *Kickoff meetings*: Where the participants are introduced, the project goals are stated and the project's direction is determined.
- *Planning meetings*: Where roles and responsibilities are assigned, the project plan is developed and decisions are made as to how the project will be proceeding.
- *Progress review meetings*: Where project principals are updated on the status of a project and near-term actions are planned and scheduled.
- *Problem solving sessions*: Where problems and conflicts are discussed with members of the team or management to arrive at and implement solutions.
- *Briefing meetings*: Where project status and progress information is shared with upper management.
- *Formal presentations*: Where project managers convey project information (such as project milestones) to managers, stakeholders, vendors and clients in a group setting.

A well-managed project meeting provides direction and clarification, addresses priorities, and keeps everyone updated on the project progress and status. The project manager should create and distribute an agenda in advance to the attendees whenever possible. Creating an agenda forces the PM to think through what will be covered at the meeting and give the attendees a heads up to help them prepare for the meeting. A well thought out agenda reduces the amount of wasted time in a project meeting and limits the conversation to topics on the agenda. Circulate the agenda in advance. Include any information that needs to be reviewed ahead of time for discussion and decision making. Recurring meetings should be scheduled at regular times and frequencies to enable attendees to keep their calendars cleared for the occasion. Keeping the meetings short will make the information meaningful and the attendees will appreciate the brevity. Meetings should be relevant and timely. Don't let the meeting get diverted from the agenda into areas of limited interest to your team members. A facilitator should be assigned to assure that the meeting stays on focus and meets its time constraints. The project manager should make every effort to cover current issues, priorities and goals, while acknowledging successes, addressing problem areas, and reinforcing the rules.

Meetings held with upper management or with people who do not report to you can get out of control if you let them take over the meeting. To avoid that scenario, establish clear procedures before beginning the meeting. Let them know that your capacity

Figure 6-2. Who's in Charge of the Meeting?

is to chair the meeting and that theirs is to make the decisions. Every meeting needs a purpose; as chairperson, you will establish the meeting's purpose and objectives. Start and end on time. If someone is late for a meeting, begin without him or her. Keep the group focused on the purpose, objectives and agenda. Ensure that everyone's ideas are heard. As the meeting ends, review the decisions and action plans to ensure they will be carried out by the person assigned and by the date agreed upon. Have the minutes of the meeting typed and distributed in a timely fashion to all attendees and others as appropriate. Meeting minutes should be brief and to the point. They should record the decisions that the group makes, an action plan of what will be done by whom, and when and how these activities will be measured. Some basic guidelines for conducting successful meetings:

- Prepare and distribute a comprehensive agenda.
- Set ground rules and post them in the meeting room.
- Explain the expectations you have of the group.
- Assign a timekeeper to keep the meeting on track.
- Assign a recorder to take minutes.
- Review the minutes from the prior meeting.
- Maintain focus on the meeting's purpose and objectives.
- Allow questions to be asked for clarification.
- Give people your opinion on the subject matter.
- Encourage all attendees to participate.
- Discourage people from changing the subject.
- Keep the group focused on the facts.
- Solicit input from each of the members.
- Analyze all suggestions and proposals.
- Weigh the positive and negative solutions.
- Let the group know where you stand on issues.
- Summarize the meeting's salient points.
- Consider the consequences of possible decisions.
- Have the group agree upon the actions to take.
- Evaluate how successful the meeting was.
- Discuss how following meetings can be improved.

- Have the minutes typed and distributed.

TELLING IT LIKE IT IS (*REPORTS*)

Time is a project manager's most valuable resource. Sometimes there don't seem to be enough hours in a day to get that day's work accomplished. Subsequently, PMs must find ways to "work smarter instead of harder!" Adequate and accurate communication of a project's status to the right people, at the right times, can have a positive effect on their use of that volatile asset. Reports focus on the transmission of timely, clearly structured and factual information to specific audiences for their review and feedback to the project manager. Depending on the audience, reports can cover the status of individual facets of the project as stand-alone documents (such as the detailing of finances, risks, resources, issues and executive summaries) or form part of a MASTER project status report. Every report takes time to assemble, time to maintain, time to read and time to act upon. If reports are to be effective conservators of time, they must be tailored to the people who are going to read them, kept as brief as possible while still informing the audience, focus on delivering the project within the project's constraints, and submitted only as frequently as they are needed. The project manager must understand what each of the stakeholders needs to know and when they need to know it.

As you move forward in your career, taking on more complex projects and become more software savvy, you'll learn about and utilize many of their features to construct and convey the above-mentioned standalone reports (in the form of risk registers, issue logs, change request logs, etc.). For now, I'm going to stick with completing an all-encompassing project status report. It's the most common and frequently used type of project report (the frequency dependent on where you are in the project and how much there is to say about it). Reporting the status of a project is a core management activity. It keeps your team on point and upper management informed of your progress. It should detail the current health of the project but avoid getting down into the weeds as to how it got there. The report should cite:

- Where things currently stand with the project.
- Any obstacles that need to be overcome.
- The project's key metrics.
- Any changes to the baseline plan.
- The next steps for moving forward.

A well thought out report projects a professional attitude and makes for easy reading by the audience. The report can be used to communicate the status of the project's schedule, budget, staffing, risks, deliverables, issues and changes. With it you can document and regularly inform stakeholders of the project's progress, raise risk and issue concerns, and ensure that all project successes are clearly communicated. At the beginning you should have a brief summary of the report content so that readers who are unfamiliar with your project can grasp what it's all about. The contents page should list the main chapters and subsections of your report. A non-technical introduction should begin with a clear statement of the nature and scope of the project and provide a summary of everything you set out to achieve. In the body of the report, reflect on the chronological development of the project, and expound on any interesting problems, features or implementations and evaluate the strengths and weaknesses of what has been done. Conclude with a listing of lessons learned and actions you have taken as a result.

THE PROJECT FILE CABINET (*RECORDS*)

A record is a hard copy document or electronically stored data residing in a project file system which acts as a guideline for project activity, or evidence of actions taken. Record management is the systematic planning, organization, tracking, storage and retrieval of that information during the execution of a project, and is an archived (historical) resource after its completion. The records outline relevant details of a project that are used to meet the project's objectives, costs and deadlines. Once it has been determined what information is required to be recorded, a systematically implemented system should have a process for handling, recording, storing, and protecting those records. And time periods should be established for their retention and disposal. Guidance should also be provided that outlines required procedural compliance and auditing requirements. For subject matter containing large volumes of information, storage and tracking of data can be accomplished through the utilization of a binder system. In such cases, each binder should be labeled with the project number, year, title and site.

Note: According to Prince Engineering, PLC, of Traverse City, Michigan, the idea for the project management file documentation system was sparked back in 1995 by a list developed by author Andrew M. Civitello, Jr., in his book *Construction Operations Manual of Policies and Procedures,* second edition. Prince Engineering has since developed its own filing system which they freely share with anyone needing such a system to aid them in their projects, stating that "we think it's too good not to share." Figure 6-3 lays out their arrangement for entering folders into a file cabinet setup.

Although I personally choose to alphabetize my folders (for quick access), their approach makes sense from the standpoint that the order they chose somewhat follows the order of a construction project: first the owner authorization or funding, then the plans and specifications, then the contract for construction and so on. How you order your folders boils down to your individual preference; either way the filing cabinet is organized. And because of its simple, repeatable fashion, it is easily learned and teaches the novice good, solid project management skills.

Their system was first used with hard copy documents, but can be used for electronic documents as well or even as tabs in three-ring binders, where project management documentation is categorized and recorded; enabling quick and easy retrieval of the information. The system keeps track of consultant, contractor and vendor responsibilities, deliverables and deadlines. The system is simple, easy to learn and very effective at recording the design, construction, and change sequences that occur during a project.

01 Owner Authorization	8/3/2017 10:13 AM	File folder
02 Bidding & Contract Documents	8/3/2017 10:13 AM	File folder
03 Plans	8/3/2017 10:13 AM	File folder
04 Spedifications	8/3/2017 10:13 AM	File folder
05 Addenda	8/3/2017 10:13 AM	File folder
06 Submittals	8/3/2017 10:13 AM	File folder
07 Shop Drawings	8/3/2017 10:13 AM	File folder
08 Bonds	8/3/2017 10:13 AM	File folder
09 Insurance Certificates	8/3/2017 10:13 AM	File folder
10 Design Clarifications	8/3/2017 10:13 AM	File folder
11 Baseline Schedule	8/3/2017 10:13 AM	File folder
12 Schedule Revisions	8/3/2017 10:13 AM	File folder
13 Permits	8/3/2017 10:13 AM	File folder
14 Meeting Minutes	8/3/2017 10:13 AM	File folder
15 Payment Requests	8/3/2017 10:13 AM	File folder
16 Correspondence	8/3/2017 10:13 AM	File folder
17 Progress Reports	8/3/2017 10:13 AM	File folder
18 Progress Photographs	8/3/2017 10:13 AM	File folder
19 Change Orders	8/3/2017 10:13 AM	File folder
20 Substantial Completion	8/3/2017 10:13 AM	File folder
21 Punch List - Final Completion	8/3/2017 10:13 AM	File folder
22 Certificate of Occupancy	8/3/2017 10:13 AM	File folder
23 Maintenance & Operating Manuals	8/3/2017 10:13 AM	File folder
24 Guarantees - Warrantees	8/3/2017 10:13 AM	File folder
25 As-Built Documents	8/3/2017 10:13 AM	File folder

Figure 6-3. The Prince Engineering Construction Folder Array
(Courtesy of Prince Engineering, PLC of Traverse City, Michigan)

To use the system, each individual contract between the owner and all designers, consultants, contractors and vendors is treated as its own sub-project. The system organizes the project by each vendor contract which then defines the tasks under the contract. Each sub-project is broken down into common categories, and project documentation is recorded in each category. For every construction project, each contractor or vendor contract is given its own scope of work, schedule and payment requirements as identified in the agreement, providing a detailed and permanent record of the construction project. Prince points out that not all contracts or purchase orders on a construction project will have documentation that fits all 25 categories. For example, no permits are required relative to furniture vendor purchase orders, but a copy of the electrical permit should certainly be placed in the electrical contractor's sub-project file.

The system is best created by entering each folder as people are hired for the project. Usually the first people hired are the architects and engineers followed by consultants, contractors and suppliers. Depending on the scope of work, each contract gets a file tree that includes the appropriate folders applicable to that contract. Not all contracts will include all 25 categories; and some may include others, such as estimates, inspections, labor compliance, legal documents, public relations and requests for information (RFIs), depending on your preference for their inclusion. A more thorough understanding of the system can be had by visiting their website at www.build-on-prince.com.

Chapter 7

Dealing with the Stakeholders

IN CHAPTER 3, I introduced stakeholders as one of the four individuals or groups of people (key players) who comprise the authority and oversight of most projects, stating that they have a vested interest (stake) in their projects and are, in essence, anyone who can affect or be affected by the project. In Chapter 4, I described the stakeholder management plan as a subsidiary plan of the project management plan and stated how any stakeholder could identify the need for a change on a project by submitting a change request form to the project manager. In Chapter 6, I explored how stakeholders can and need to communicate with the project manager and with one another. Obviously, stakeholders hold a lot of sway and are very involved in the planning, execution, monitoring and control of projects.

Although the Project Management Institute (PMI) covered stakeholder activity in their prior editions of the PMBOK (*Project Management Body of Knowledge*), treatment of the subject was interspersed throughout the guide, similar to the way I am referring to it in this book. It wasn't until the current (5th) edition, that the subject was given its own stand-alone attention, when PMI expanded the nine knowledge areas to ten; the tenth being *project stakeholder management*. According to the PMBOK guide, project stakeholder management includes the processes required to identify the people, groups or organizations that could impact or be impacted by the project, to analyze stakeholder expectations and their impact on the project, and to develop appropriate management strategies for effectively engaging stakeholders in project decisions and execution. It goes on to explain that stakeholder management focuses on continuous communication with stakeholders to understand their needs and expectations, addressing issues as they occur, managing conflicting interests and fostering appropriate stakeholder engagement in project decisions and activities. The knowledge area covers four project stakeholder management processes:

Identification of Stakeholders—(*Initiation Phase*)
Identify, analyze and document information regarding stakeholder interests, influence, involvement, and potential impact on project success.

Planning of Stakeholder Management—(*Planning Phase*)
Based on the analysis, develop strategies to engage the stakeholders throughout the project life cycle

Management of Stakeholder Engagement—(*Execution Phase*)
Work with stakeholders to meet their needs, address their issues and engage them in project activities.

Control of Stakeholder Engagement—(*Monitoring and Control Phase*)
Monitor overall project stakeholder relationships and adjust existing strategies and plans.

UNDERSTANDING AND IDENTIFYING PROJECT STAKEHOLDERS

The management of project stakeholders is critical to the success of every project. It is the PM's responsibility to educate stakeholders on what is realistic and what to expect from a project. The management of stakeholder expectations can be difficult because of conflicting goals and expectations. Subsequently, project managers must have a keen appreciation for the expectations and motivations of their key stakeholders. Once a project is initiated, those key project stakeholders need to be identified, and then a plan must be formulated for managing their individual needs and assuring that their expectations are realistic. Only then can the PM discuss the project with them and get more specific about its goals and deliverables.

Stakeholders can be internal or external to an organization. Internal stakeholders are those directly affected by the project, such as staff members. External stakeholders are interested parties that are not a part of the business, such as vendors, suppliers or lenders. They may exhibit high or low powers of authority and can exert positive or negative influences on a project, its deliverables, and its team in satisfying their own agendas. Positive stakeholders see the project's beneficial side and aid the project management team to successfully complete the project. Negative stakeholders see the downside of the project and are less likely to add to the project's success.

As mentioned, stakeholders are typically the project sponsors, members of a project team, project managers, executives, management personnel, the media, contractors, sub-contractors, consultants, community and government agencies, banks, suppliers, customers, clients, and/or end users of a project's product or service. Key stakeholders will have to be involved early on to get their input for the project overview, goals and deliverables. To identify stakeholders that you may not have recognized as such or at first, you may want to ask known stakeholders of their awareness of others who might have an interest in the project and who haven't already been identified. Leaving out an important stakeholder could have an adverse effect on the project. The sooner all of the stakeholders are identified, the sooner you can communicate with and involve them in the project activity. Effective communication between stakeholders ensures that everyone is on the same page and that the project will continue to move forward.

It's important for the project manager to know the stakeholders' interests and understand their roles in the project and the organization. They help decide on issues from the beginning, during planning and at execution of the project. PMs should establish and nurture good working relationships with them, to learn about their business concerns and needs, fostering easier handling of pressing issues down the road. Positive (proactive) stakeholders help the project team develop the best possible strategies. It's important to identify positive relationships between them. Those relationships build trust and encourage collaboration. Negative (adverse) stakeholders can adversely affect a project's outcome. Failing to recognize adverse stakeholders can discourage collaboration and hinder project development. Identify the difficult people, and keep an eye on them. Keep your ears open, as well; listen to what they're saying, and try to understand their motivation and goals. Try looking at things from their point of view, and determine if there is room for compromise.

When deciding who to consider as a stakeholder for the project, ask if the person or organization can be directly or indirectly affected by, or is in a position to influence the project. Can they impact its funding, personnel or material resources? Are their skill sets or capabilities required by the project? And are potential stakeholders dependent on one another aside from the project? Once chosen, classify them as primary or secondary. Primary stakeholders (customers and end users) have a major interest in the success of a project because they are directly affected by its outcome. Secondary stakeholders (administrative, financial and legal), though not directly affected, assist with processes, aiding in project completion. Determine whether they are high or low power. High-power people (upper management and end users) can influence project decisions, and must be fully engaged, while low-power folks (everyone else) need only to be kept updated on a project's status.

There are 47 total processes that can be found in the PMBOK® Guide Fifth Edition. Identifying stakeholders is the first of the four processes that comprise the project stakeholder management knowledge section. Each of the processes is designed to produce some output by applying selective tools and techniques to a set of inputs. Those ITTO (inputs, tools and techniques, outputs) components form the basis for executing projects.

For instance, to develop a plan for a given project before its inception, you would need to determine what items (in-

Inputs	Tools and Techniques	Outputs
Project Charter	Stakeholder Analysis	Stakeholder Register
Procurement Documents	Expert Judgment	
Enterprise Environmental Factors	Meetings	
Organizational Process Assets		

Figure 7-1. Identify Stakeholders ITTO

puts) would be required to facilitate its creation. To develop a stakeholder register, you could use templates that currently exist within your organization (organization process assets). To identify stakeholders, you can draw from the organization's cultural norms (enterprise environmental factors). And the initial project requirements details can be gleaned from the project charter. Those inputs provide support for your planning efforts and structure. The tools and techniques then aid you in utilizing those inputs by working with the identified stakeholders, whose derived opinions culminate in the creation of the stakeholder register. In the PMBOK® Guide Fifth Edition, the output (from one process can be an input for another. In this case the stakeholder register (output) from the IDENTIFY STAKEHOLDERS process is an (input) to the PLAN STAKEHOLDER MANAGEMENT process, within the stakeholder management knowledge area. The stakeholder register can be found as an input in five other knowledge areas, as well; specifically, the procurement management, risk management, communications management, scope management and quality management knowledge areas.

PLANNING TO MANAGE THE STAKEHOLDERS

Having identified the project stakeholders (now in the stakeholder registry) during the initiating phase, the project manager will need to develop a management strategy for effective stakeholder engagement. Based on the analysis of their interests, needs, and potential impact on project success, the (plan stakeholder management) process involves identifying and developing management strategies and mechanisms that will achieve the greatest support of the stakeholders and effectively engage them throughout the life cycle of a project.

The tools and techniques to be utilized will include expert judgment (of the PM and other stakeholders), the use of analytical techniques (such as a stakeholder assessment matrix used to track stakeholder engagement and attitudes) and regularly scheduled meetings for stakeholder input, discussions and feedback. Outputs are the stakeholder management plan (a component of the project management plan), that will cite strategies for managing the stakeholders and show their relationships and communications requirements. That plan will also state the desired and current levels of engagement and list the format of the information to be distributed. The key benefit of this process is to provide a clear, actionable plan of interaction with stakeholders in support of the project.

Note that the stakeholder management plan is a sensitive document and should be protected from unauthorized disclosure of its contents. Having identified the project stakeholders (now in the stakeholder registry) during the initiating phase, the project manager will need to develop a management strategy for effective stakeholder engagement. Based on the analysis of their interests, needs, and potential impact on project success, the (plan stakeholder management) process involves identifying and developing management strategies and mechanisms that will achieve the greatest support of the stakeholders and effectively engage them throughout the life cycle of a project.

MANAGING THE ENGAGEMENT OF THE STAKEHOLDERS

The stakeholder registry explicitly describes the individuals, agencies and institutions who share a stake or an interest in the project. The relevancy of those principals and their opinions

Inputs	Tools and Techniques	Outputs
Project Management Plan	Expert Judgement	Stakeholder Management Plan
Stakeholder Register	Analytical Techniques	Project Document Updates
Enterprise Environmental Factors	Meetings	
Organizational Process Assets		

Figure 7-2. Plan Stakeholders Management ITTO

cannot be overstated, and the impact of their decisions must be considered in any rational approach to the management of a project. Besides identifying and assessing the impact of stakeholders who are subject to the authority of the project manager, the PM must consider how the project's goals and objectives will affect or be affected by stakeholders outside of their authority. Projects are based upon technological, political, social-economic, legal, and myriad other systems that can influence an organization's growth and survival. Their potential impact calls for the project manager to continually engage all stakeholders (both internal and external), not only to ensure the successful outcome of their project, but to avert any adverse organizational consequences that they may pose. This can be accomplished by following the strategies laid out in the stakeholder management plan. In it is described a process to foster appropriate stakeholder engagement for communicating and working with stakeholders to meet their needs and expectations and address issues as they are manifested. The strategies are designed to increase support for and minimize resistance from stakeholders. Actions include:

- Providing a forum for discussing topics relevant to the project.
- Anticipating problems, addressing concerns and resolving issues.
- Confirming stakeholder commitment to the project at each of the five project stages through continual communication.
- Ensuring project goals are met by managing expectations through engagement and negotiation.

Active management of stakeholder involvement decreases the risk of project failure. The actions will encourage active support for the project, minimize negative impacts and culminate in better project decisions. Used in conjunction with a communication management plan, the SMP provides information and guidance to the stakeholders throughout the project's life.

Inputs	Tools and Techniques	Outputs
Stakeholder Management Plan	Communications Methods	Issue log
Communications Management Plan	Interpersonal Skills	Change Requests
Change Log	Management Skills	Project Management Plan Updates
Organizational Process Assets		Project Document Updates
		Organizational Process Assets updates

Figure 7-3. Manage Stakeholders Engagement ITTO

MONITORING AND CONTROLLING PROJECT STAKEHOLDERS

Control stakeholder engagement is the process of monitoring and controlling project stakeholder relationships by adjusting the plans and strategies used for engaging them. Again, the objective, as with all the other 46 processes, is to produce outputs using the stated tools and techniques to manipulate the information gleaned from the available input information. In this case, the outputs track work performance (the effectiveness of the stakeholder engagement process) establish change order requests (action needed to bring engagement into alignment with the plan), and result in updated documents, plans and processes (such as the stakeholder register and issue log). Monitoring and controlling stakeholder engagement is the action of tracking the work being performed, determining the adequacy, accuracy and appropriateness of it and surmising whether it will accomplish the end goals of the project successfully. Stakeholders can be supportive of or detrimental to a project. To manage either, it's imperative that you develop implementation strategies for dealing with them. Creating an organizational policy, stipulating that stakeholders will be actively managed is important. Once in play, additional action plans can be developed for dealing with them. The additional policies and procedures can be constructed to ensure that:

- Stakeholders fully appreciate their potential impact.
- Stakeholders respond to major project decisions.
- Project review meetings are scheduled and facilitated.
- Stakeholder assessment is integral to determining project status.
- External stakeholders maintain contact with the PM.
- Ongoing status reports are kept up to date.

Inputs	Tools and Techniques	Outputs
Project Management Plan	Information Management systems	Work Performance Information
Issue Log	Expert Judgement	Change Requests
Work Performance Data	Meetings	Project Management Plan Updates
Project Documents		Project Document Updates
		Organizational Process Assets updates

Figure 7-4. Control Stakeholders Engagement ITTO

Chapter 8

Issues, Conflict, Problems and Risks

DEPENDING ON HOW THEY ARE USED, these four words can have similar connotations. Look them up in a thesaurus, and the terms may at times seem interchangeable. But in reality, at least from a project management perspective, their meanings are very divergent and quite specific. Never the less, if not properly managed, any one of the four can spell "TROUBLE" for a project. Their commonality comes from the fact that each of them can be found in every project. Here's my shot at delineating their differences:

ISSUES are *controversial* problems encountered when executing project activities (i.e., problems with staff members, technical failures, material shortages or schedule aberrations). An issue is something that has already occurred (essentially a risk that has happened) or a problem that exists, which can impair a project's successful completion. Generally speaking, they are problems that need to be solved when they are manifested. Unresolved issues can result in conflicts, unnecessary delays, and failure to produce project deliverables.

CONFLICTS can be defined as clashes or disagreements, between two or more opposing individuals or groups, having different objectives and attitudes. They arise when individuals from different backgrounds and orientations are called upon to work together to complete complex tasks. Interpersonal conflict occurs when people harbor unspoken assumptions, disagree with objectives, decisions and actions, or exhibit any number of human emotions (i.e., fear, anxiety, stress, envy, anger, etc.). Like issues, conflicts are problematic and need to be resolved. A small problem can turn into a huge one if conflict is allowed to fester. Unresolved conflicts lower group morale, undermine team harmony and can result in failed goals.

PROBLEMS are negative difficulties or complicated situations that need to be resolved (something bad that has to be dealt with). They come in different magnitudes (major and minor significance) and can adversely impact projects and their stakeholders. Problems are regarded as unwelcome, harmful, or just plain wrong, requiring resolution of the existing state in deference to the desired state. Unlike issues, problems are not controversial (see issues above); they are more straightforward and emphatic (a malfunctioning crane or failed lighting string, for example). Problems tend to be less predictable than issues, and can arise with little or no warning. Most problems are handled as they occur, with only a mention of their resolution at the progress meeting (to capture costs, time spent and actions taken) avoiding the need for a log or register to track them.

RISKS are defined as events that are anticipated as possible, but that have not yet happened (potential problems). They are unknown events that might (or could) take place during the course of a project which could affect its cost, schedule, scope or quality. Poor risk management can lead to project failure. It is important to identify risks before beginning a project and identify and track them in a risk register as the project evolves. Project risk is covered extensively in the PROJECT RISK MANAGEMENT knowledge area in the PMBOK guide, 5th edition.

MANAGING PROJECT *ISSUES*

Managing projects is a complex undertaking. They involve many variables, including people, money, work assignments and concerns (to name a few). To cope with it all, savvy project managers use various tools and methodologies to effectively communicate and manage their projects. One such tool is documentation, which is utilized for recording, processing and communicating project information. Issue logs are used to capture project concerns, communicate them to the project team, and assign responsibility for their resolution. When issues go unresolved, adverse situations materialize that can negatively impact a project. Issue management is the process of identifying and resolving those situations before that happens. The purpose of the issue management process is to assist the project team with identifying, analyzing, validating, and resolving the issues impacting a project and to communicate their status to the project stakeholders.

The process begins with the recognition of a concern levied by a stakeholder to the project manager. Once recognized, a general description of the issue should be created that documents its potential impact and provides a suggested solution. The project manager must then decide whether the issue should be handled formally and tracked, or informally by noting it in the minutes of the progress meeting. If the issue is to be handled formally, then the full details will need to be recorded in the issue log to track the sender and date of the submission. The issue should then be assigned to an individual or group who can most likely resolve it. That individual or group must then investigate the subject, resolve the issue and report the action back to the project manager with a description of the resolution. The project manager then officially closes the issue.

Figure 8-1 illustrates two issue logs examples; Figure 8-1b is an excerpt of the actual log I created for managing the issues of a facilities management pilot project I led at a major medical facility compound. Feel free to modify and use it for your own needs. Alternatively, you can choose to create your own log to reflect the needs of any project you may oversee or adopt one of the templates that have graciously been made available for free by a number of organizations on the internet. Figure 8-1a is one such document furnished by Project Management Docs. Depending on your circumstances and whatever issue log you choose to incorporate, here are some components that may be appropriate for inclusion:

- Issue reference number (number of each issue).
- Issue's name (issue ID and description).
- Issue's author (the person who raised the issue).
- Parties involved (the people involved with the issue).
- Issue type (the category it falls into).
- Issue priority (the order in which it will be addressed).
- Issue severity (the seriousness of the issue).
- Issue date (when the issue is raised).
- Date assigned (when the issue is assigned).
- Deadline (the planned settlement date).
- Issue update (actions taken before resolution).
- Current status (open, in progress, closed).
- Date resolved (when it's actually completed).
- Resolution (how the issue was resolved).

Some organizations (and projects) call for strict compliance with established rules or the use of predetermined templates for monitoring and controlling these four subjects; but, for the most part, the means utilized for tracking them is left up to the project manager. Due to the nature of the facilities management pilot project I lead, I designed the above form (Figure 8-1b), to address both ***issues*** and ***problems***. ***Conflicts*** were handled extemporaneously, either face-to-face with the combatants, or in meetings calling for the express resolution of the difficulties encountered. ***Risks***, on the other hand were relegated to the process used for that knowledge area in the PMBOK. The 44-page Project Risk Management section provides an overview of the risk management process, defines how the outcomes of the process will be documented, analyzed, and communicated, and describes the content and format of the *risk register*. **Author's note**: The risk register was our preferred choice for dealing with risks associated with the facilities management pilot project.

Issue Log								
Project:							Date:	
Issue	**Description**	**Priority (H,M,L)**	**Category**	**Reported By**	**Assigned To**	**Status**	**Date Resolved**	**Resolution/ Comments**
This should be a standard numbering system.	Detailed description of the issue.	High, Medium or Low priority.	Assign to a category.	Who reported the issue?	Who is the issue assigned to?	What is the status of the issue?	What date was the issue resolved?	What was the resolution or what is being done to resolve the issue?

Figure 8a: Free Issue Log sample from the internet

		Facilities Management Pilot					
ARCHIVE		Kenneth Petrocelly					
CLOSED ISSUES		2/29/2012	ISSUES and ACTION ITEMS				
BUILDING	Issue / Action Name	Description	Status	Open Date	Closed Date	Currently Assigned	Comments
A	PDA Routes	Building A routes to be download	Closed	2/1/2012	na	Contractor	Whatever they have to
14A	PDA Routes	Building A routes to be download	Closed	2/1/2012	2/3/2012	Contractor	Additional routes to be
14A	Handheald readings	Walk with contractor on route 14a	Closed	2/1/2012	2/7/2012	Contractor	Scheduled for (9 AM)
All	Belt alignment kit	Belt alignment kits forthcoming ro	Closed	2/8/2012	2/15/2012	Contractor	Will deliver to Martin S
All	AAALAC work	List of AAALAC jobs planned and	Closed	2/8/2012	2/15/2012	Dave C	Dave will provide a list to Martin S
All	Building Drawings	Burn CDs for Building(s) 14 group	Closed	2/8/2012	NA	Wayne S	Distribute CDs
N/A	Weekly Status Meeting	Pass meeting 'Baton' to Wayne Sh	Closed	2/8/2012	2/22/2012	Ken P / Wayne S	Turn over & explain document forms
ALL	Equipment inventory	Parent / child configuration choic	Closed	2/8/2012	2/22/2012	Martin S / Barnard	Sarge to confer with
14D	Poor heat	Need new converter for 100 area	Closed	2/15/2012	2/22/2012	Martin S / Barnard	Check with contractor
28	Mechanical areas of bu	Review CDs	Closed	2/15/2012	N/A	Ken H	Review information
All	AAALAC Report	Areas to be repaired to AAALAC C	Closed	2/22/2012	N/A	Mark W	Received report from Dave W

Figure 8b Weekly meeting snapshot of Facility pilot issues

Figure 8-1. Issue Log Examples

MANAGING PROJECT *CONFLICTS*

As you become more attuned to PMIs PMBOK guide and seek it out for your direction, you'll both welcome and appreciate the guidance it provides. In it are 64 references to conflicts and their management, interspersed throughout its text, with a heavy emphasis on human resource management. Conflicts at best can cause dysfunction within a project; at worst, they can completely sabotage it. Managing conflict is an essential attribute and one of the biggest challenges faced by project managers, as it draws on a cadre of interpersonal skills. Project managers must be able to identify their causes and actively manage them, to minimize their potential to negatively impact their projects. If actively managed, the project team can then deliver better decisions, thereby increasing the probability of their project's success. Answers to the situations we deal with are hardly ever black or white; inevitably the resolutions are almost always found exploring the gray areas. But, using specific management techniques or approaches for those situations invariably culminates in specific results. Some available instruments include the Thomas-Kilmann Instrument (TKI), the Blake Mouton Managerial Model Adaptation to Conflict Resolution, Rahim's meta-mode and (of course) PMIs PMBOK guide. The PMBOK guide states the obvious fact that "conflict is inevitable in a project environment"; ergo, they provided five techniques for managing conflict. Here they are (in no particular order):

Withdraw/Avoid—(I lose, you lose)
[If there's no fight, there's no winner]

But, sometimes withdrawing (the act of avoiding the conflict all together) can be an effective strategy, such as when the matter being debated is too small a fish to fry in the big project pond. The decision to drop the item may reduce the stress levels of the conflicting parties, thereby lowering the tension in the group; and postponing the issue may provide time for a more thorough investigation or allow it to be resolved by others. One caveat—if you choose that tact and fail to follow up, the magnitude of the conflict could be compounded.

Smooth/Accommodate—(I lose, you win)
[Giving in to the other side]

According to the PMBOK, the act of accommodating in conflict management is "conceding one's position to the needs of others to maintain harmony and relationships." The technique calls for dealing with the concerns of others, in deference to your own. It requires understanding of the conflicting parties and is best applied when one side has more to gain (or lose) than the other. Deferring to the judgment and opinions of others can result in improved stakeholder relationships, but often at the cost of your being perceived as too accommodating, particularly when dealing with pugnacious or aggressive individuals or groups.

Compromise/Reconcile—(I win some, you win some)
[I'll scratch your back, if you scratch mine]

Recognizing that some conflicts cannot be fully solved, conflicted parties sometimes search for middle ground that they can feel comfortable with. Compromising emphasizes areas of agreement rather than areas of difference, and involves getting two or more parties (having divergent opinions) to agree upon a mutually satisfactory end result. It involves discussions between the parties; each giving in on simple but specific ideas or suggestions by the other side, to move things along. In essence, both parties capitulate, searching for solutions that bring some degree of satisfaction to all parties involved in making their decisions. The success of utilizing the technique hinges on two things: how well the project manager understands the stakeholder's needs, and the requirement for all parties to benefit from the agreements they arrive at. The compromise approach is generally not adversarial, is quickly rendered and can act as a judicious stand-in until better, more permanent decisions can be explored. To be effective, all parties need to be open and honest, and the agreed upon actions should be monitored for compliance.

Force/Direct—(I win, you lose)
[My way or the highway]

An alternative take on this technique might be "No more mister nice guy!" As a project manager, forcing action on any matter, as a consequence of your authority to resolve conflicts, can come at a cost. Forcing or directing is sometimes referred to as competing, as when one party pushes their idea at the expense of another (such as when disagreements arise over design features). Most likely the harm suffered will be in the form of breakdowns in the relationships that project managers have with their project teams. But, from time-to-time, project managers need to take a stand and apply their power; for instance, when they require their members to adhere to OSHA requirements, or stop a project when they witness a dangerous situation. Whereas its application can provide a quick resolution to a conflict, it can often lead to undesirable, bitter and heated exchanges. Obviously, the force/direct conflict management technique should be used only sparingly. **Note**: As a facility manager, I once stopped a project dead in its tracks when I witnessed repairmen walking over a (non-supported) makeshift, 2" x 8" wooden plank to access a cooling tower fan gear box, 20 feet above the metal water catch basin (see Figure 8-2). The situation was ultimately resolved with the construction of a scaffolding system and creation of a repair work plan.

Collaborate/Problem Solve—(I win, you win)
[United we stand…]

The collaboration and problem solving technique appears to be the most utilized approach to resolving project conflicts. Its use requires conflicted parties to foster an open dialogue regarding their different viewpoints and perspectives. Unlike compromising, collaborating may take more time and energy to render decisions but generally results in more permanent, mutual agreements that are satisfactory to all involved parties. The benefits are that the conflicts themselves are permanently resolved, and the project team is strengthened as a consequence of their positive interaction. The process calls for all parties to bring their

Figure 8-2. The scaffolding was constructed inside the chambers.

valid, defensible, salient points to the discussion, and then work through them with one another until a consensus is arrived at and implemented. The approach averts further conflicts downstream, promotes improved trust between team members and re-enforces the collaborative atmosphere.

MANAGING PROJECT *PROBLEMS*

In the project management arena, issues and problems are "kissing cousins"; whereas issues are associated with conditions and complications, problems are considered negative obstacles that need to be solved. Problems may involve people, matters or situations that are difficult and perplexing. Small problems can evolve into big ones if there is conflict and it's allowed to fester. All problems need to be identified, defined, have their causes determined, corrective actions prioritized and solutions implemented. Problem-solving strategies are the steps taken to make that happen and should be a component part of your overall project.

Team members should be encouraged to look for potential problems and report problems when they occur. Collaborative efforts by project team members will ensure that problems are dealt with promptly, successfully and permanently. When necessary, meetings should be held (attended by relevant parties) to address any problems that threaten the project. The agenda should be clearly defined, with established ground rules for staying on topic, and minutes should be recorded to document the discussion and actions agreed upon.

Applying problem solving techniques, team members can quickly identify problems and collaborate on their solutions. Project managers should explain the importance of bringing problems forward, construct a structured step-by-step method for solving them and encourage team members to express their concerns and offer solutions. The first (and most important) step in that process is to identify a problem and recognize its root cause, so that it can be clearly understood. Once completed, a discussion should ensue—addressing the possible impact the problem might have on the project—followed by an exploration of all available solutions. After ample consideration of all the possibilities, an approach can then be selected for solving the problem, and its implementation can be planned and executed. After solving the problem, the team should verify the results, and make certain that the resolution does not create any additional issues. They should then correct the root cause of the problem, so that it isn't repeated. That said, the following are some techniques that can be used for setting up a strategy.

Brainstorming—letting the team suggest a large number of ideas, then developing them until a viable solution is agreed upon.

Data collection—making sure relevant information is collected, team members can be more likely to develop suitable responses.

Intuitive reasoning—approaching solutions indirectly and creatively.

Modeling—first solving the problem in a model system before its application in yours.

Patience—use a patient, methodical approach on the matter to help avoid any mistakes.

Research—employing existing solutions to similar problems.

Root cause analysis—identifying the cause of a problem.

Supposition—assuming a possible solution for the problem and trying to prove or disprove the assumption.

Trial and error—trying possible solutions until you find the one that works.

MANAGING PROJECT RISKS

Like everything else in life, projects don't always go as planned. Every project is replete with components of uncertainty (risk), and conscientious management of those components must be made an integral part of assuring their success. Risk management is concerned with how those uncertainties are dealt with. Like project stakeholder management, project risk management is such a large part of the project management process, that it warrants its own knowledge section in the PMBOK. In it, risk management contains the processes for identifying, analyzing and responding to project risk. And like the stakeholder section, each process has a set of inputs, a set of outputs and a set of tools and techniques that are used to turn those inputs into those outputs. The six processes involved, as detailed in the PMBOK, are:

Plan Risk Management—defining how to conduct risk management activities for a project.

Identify Risks—determining which risks may affect the project and documenting their characteristics.

Perform Qualitative Risk Analysis—prioritizing risks for further analysis or action by assessing and combining their probability of occurrence and impact.

Perform Quantitative Risk Analysis—numerically analyzing the effect of identified risks on overall project objectives.

Plan Risk Responses—developing options and actions to enhance opportunities and to reduce threats to project objectives.

Control Risks—implementing risk response plans, tracking identified risks, monitoring residual risks, identifying new risks, and evaluating risk process effectiveness throughout the project.

Risk and risk management is an involved and complex area of study, and its total understanding is beyond the scope and purpose of this book. Even the 44-page narrative in the PMBOK (though it sheds a large light on the subject matter), doesn't tell the whole story. Naturally, it would behoove you to grasp all of the enlightenment you can from the risk section of the guide, but my suggestion is to augment that learning with separate texts, dedicated specifically to the topic.

Author's note: In addition to the coverage provided in the guide, PMI also publishes a practice standard for project risk management. (PMI practice standards are guides to the use for a tool, technique, or process identified in the PMBOK guide or other PMI standards). "The purpose of the Practice Standard for Project Risk Management is to (a) provide a standard for project management practitioners and other stakeholders that define the aspects of Project Risk Management that are recognized as good practice on most projects most of the time and (b) provide a standard that is globally applicable and consistently applied." The standard can be used by project management practitioners to validate the risk management process being employed in a specific situation, project or organization, but does not prescribe how the process should be implemented. It can be purchased from PMI, or members can download it for free as a pdf document.

Volumes have been written on the subject that fills library

shelves. Our focus in this chapter is to acquire a base understanding of risk, how it differs from the other three conditions, and to suggest a mechanism for monitoring and controlling it. That mechanism, in this case is, the *risk register*. (see Figure 8-3). It's a component part of your risk management plan and a necessary document for keeping your project unencumbered and on track. In project work, it isn't always necessary to avoid all risks, but it's imperative that you manage them. You shouldn't necessarily avoid all risks, but you do need to take steps to manage them properly. The importance of using a risk register cannot be overstated. It's used to record identified risks, the assessment of their severity, and the actions to be taken for their mitigation. The registers can span a wide range of formats (spreadsheets, tables or simple documents), and their frameworks may be populated with any number of components (risk IDs, descriptions, and types; low/medium/and high levels; severity, priority or likelihood to occur; dates of acknowledgement, status or mitigation; the owners, actions taken, etc.). Keeping registers updated throughout the duration of a project is crucial for success. As is the case with issue logs, risk registers can be home grown or acquired (gratis) from a great number of organizations via the internet. Every organization has a different approach to the risk register (log). This example includes all of the information that needs to be captured, and has been used on numerous real world projects. Feel free to tweak this to fit your project.

COMMON SENSE TROUBLESHOOTING

As you can see, trouble comes in a variety of forms, and each has its own time cycle and duration. Dealing with trouble takes both understanding its different kinds and the establishment of procedures for their elimination. Aside from formal policies and methods for their handling, there is (what I refer to as) a "common sense troubleshooting" approach that should always be your first defense when dealing with troubles encountered on

ID #	Date	Risk Description	Likelihood	Impact	Severity	Owner	Mitigating action	Contingent action	Progress	Status
1	12/12/17	There is a risk that.... If this happens.....	High/ Low Medium	High/ Low Medium	See Severity Table	Person managing the risk	Actions that can be taken to reduce the likelihood of the risk occurring. May also be acceptance of the risk or transference of the risk	What will be done if this risk does occur? Usually actions to reduce the impact on the project	Action taken and date	Open, Waiting, Closed

Sample Risk Register

Figure 8-3. Example of a risk log showing suggested column information.

a project. Using a "common sense" course of action in the early stages of uncovered trouble spots can lead to their quick and permanent conclusion. Common sense actions are analogous to the five human senses, sight, sound, smell, taste and touch. Here is how I describe it:

- Sight—If, during your frequent inspections of the project work site, there appears to be something out of place or just plain wrong, look to quickly correct the situation or condition.

- Sound—If, during your interactions with project personnel, you hear that something just isn't right, or that difficulties are being experienced, listen intently to them and act on their concerns.

- Smell—If, during your inquiries regarding the team's relationships, something "stinks to high heaven," hold a rapprochement meeting to let them air out their differences.

- Taste—If, during negotiations with your vendors and consultants, their actions leave a bad taste in your mouth, cut off the proceedings until their input is more palatable.

- Touch—If, during your observations of the stakeholders, you feel that things are going south, manage them "hands on" until everyone is more comfortable with its direction.

Part III — The Practitioner

Chapter 9

From Striving to Thriving

IN THIS CHAPTER I'm assuming a mentoring posture. My intent is to give newcomers to the craft an understanding as to where they fit in the PM hierarchy. Through my personal career overview, I'll describe the traits and characteristics that they need to possess (or acquire) if they are to evolve from novice to expert status as they proceed along their project management career path. I'll explain the steep learning curve that confronts those who aspire to a higher level of ability and expertise, and share which organizations, programs and methodologies I used to succeed in my own quest. Over the past 40 years I've held responsible positions in every facet of facilities engineering, plant operations and project management including construction, safety, security, maintenance, operations, communications and environmental services, both at the rank-and-file and managerial levels, exercising control over as many as 164 hourly and 5 direct managerial reports. I've worked in high-rise and campus style, multi-building and multiple location settings, in both union and non-union environments. As an ad hoc project manager, I oversaw managerial and industry-based projects and programs (of ever increasing complexity and responsibility), before acquiring my Project Management Professional (PMP) certification from the PMI, which enabled me to up my game and enter into the lucrative government contracting arena, where I've spent my last 21 years.

ON BECOMING A PROJECT MANAGER

The road to project management expertise is a long and winding one, with many different spurs and off-shoots. Traversing it is never a smooth and unencumbered ride. There are many ruts and pot holes you'll need to avoid if you are to finish the journey unscathed. This was the route I chose to take to arrive at that destination.

At the beginning of the book, I shared my experiences dealing with the first project I was tasked to perform. Frankly, the assignment was made as a consequence of the facility's need (where I worked), and the expectation that I should be able to complete it, in my position as the plant operations manager. Fortunately for me, at that early stage of my career, I had acquired the skills, traits and characteristics to pull it off. Looking back on the project, it obviously called for leadership qualities, a capacity for organizing, effective communications, the capability to manage a team, some engineering competency, a creative imagination, problem solving skills, number crunching skills and a talent for negotiation. From my 4 years in the U.S. Navy, the leadership, communications, team management and engineering background prepared me for work in hospital power plants through which I worked my way into operations management. I got my background in the engineering division. My number crunching and negotiation skills came from my business management classes and working on the job with contractors. And my creative imagination, capacity for organization and problem solving abilities have always been an inherent part of my personal make-up. That combination served me well on the parking lot project and formed the basis of my system for dealing with all my project endeavors.

As the number and complexity of the projects I managed grew, I acquired a more sophisticated knowledge base and diversity of capabilities to take them on. The emergency treatment suite (design) I mentioned in Chapter 4 not only introduced me to drawing software, but required me to gain a better understanding of the various utility systems, their components and interconnections. Building on that newly acquired knowledge enabled me to take on projects of ever increasing scope, and magnitude. Subsequently, I designed a 4,500-square-foot "Family Practice Residency" within the confines of the hospital I then worked in.

The installation was completed using only in-house maintenance mechanics for its construction. During that project, I learned a great deal about materials and the functional adjacencies of different department operations. Similar projects educated me to the point that I was able (at a much larger facility) to develop an in-house construction capability (no longer having to interfere with maintenance operations) at a hospital in Kentucky; to build a $4 million "Women's Pavilion" (see Figure 9-1) and complete a $3 million "Laboratory" expansion. Delving into those higher-end projects called for the enhancement of my construction activities understanding, and the methodologies used for managing projects, through attendance of classes at universities and a lot of personal research and reading of that subject matter.

Combining all of that with my hands-on and management background in plant engineering, I was able to take on different types of projects (other than construction) such as the installation of computerized maintenance management software (CMMS) programs and the creation of standard operating procedures

Figure 9-1. More like a Hotel than a Hospital

(SOP) and safety and security manuals. All of which boded well for me, qualifying me to take on free-lance project work (still without the ever having a certification).

On one such contract I managed several projects at a New Jersey based facility which included, reworking of their data center cooling, replacement of the central bulk oxygen tank, removal of an underground oil storage tank (UST) and up-grade modifications to the fire suppression system. Eventually there came a time when the projects became so large and diverse that it became paramount that I find a way of working "smarter instead of harder." Sure, I learned a lot of shortcuts and tricks along the way, experienced a great deal of achievement, and gained an appreciable amount of insight into many processes, but (if I was to move on to bigger and better undertakings) it was time to buckle down and attain the formal education I needed to become certified. That called for me to find the definitive organization that could provide me with all of the knowledge and resources I would need to establish myself as an expert in the field and expand my career horizons. I set my sights on that goal and learned everything I could about adopting a methodology, choosing a professional project management software program, determining best practices, putting together a formal project plan and filling in any gaps I had in my understanding of project management processes.

CHOOSING PROJECT MANAGEMENT SOFTWARE

Organizations don't always have standard practices for managing projects, especially those that are not project based. Often, in addition to performing their everyday duties, when managers are called upon to don the hat and mantle of a project manager, they find themselves woefully unprepared to take on the challenge. Subsequently, they scramble to arm themselves with whatever tools they can find (and quickly learn about), that can assist them in carrying out their assignments. As I stated in Chapter 3, every project has project goals and objectives which are carried out under certain constraints, the major ones being schedule (time), cost (budget) and scope. That "triple constraint" forms the basis for what project managers need to monitor and control in every project. When managing small and simple projects, of limited scope and duration, the skills and tools they already possess and utilize in their everyday activities may suffice to begin a project, execute it and bring it to a satisfactory close. Examples might be: using spreadsheets to create a timeline, using their accounting acumen to control costs, and writing all their documents and reports using word processing software. Besides those measureable constraints, project managers are confronted with countless other success criteria, such as product quality, meeting safety standards, achieving the project's purpose and providing sustainable reliability. As their projects become lengthier and more complex, they will need to use more powerful software applications which contain more features and possess a higher level of sophistication. Software features may include, but not be limited to:

- Budget management
- Calendar
- Collaboration
- Communication
- Cost control
- Departments and contacts
- Documentation
- Gantt charts
- Issue management
- IT project management
- Milestone tracking
- Percent-complete tracking
- Reporting
- Resource allocation
- Portfolio management
- Project planning
- Requirements management

- Resource management
- Scheduling
- Status tracking
- Task management
- Time and expense tracking

The features outlined above are offered by the majority of quality project management applications, it is important to understand that not all applications have all of those features. Selecting which bundle you choose for your project (and organization) will depend on your established requirements, the ease with which the system can be implemented and the acceptability of its cost. Four points to consider when choosing project software are planning abilities for mapping out an entire project from beginning to end, task management capabilities to break down complex projects into manageable work packages, sharing and collaboration functions to make documents and other project-related materials available to team members, and an acceptable pricing option. There are myriad project management software bundles available on the market (from free to pricy), possessing applications of varying ability and focus. I have used several of them during my career but use MS Project as my preferred tool for managing all of my projects' activities. Microsoft Project is a project management software product (part of the Microsoft Office family) designed to assist a project manager in developing a plan, assigning resources to tasks, tracking progress, managing the budget, and analyzing workloads.

ADOPTING A PROJECT MANAGEMENT METHODOLOGY

"There's more than one way to skin a cat" is a cruel analogy, but one often used to explain that there are alternate means available for accomplishing an end goal; i.e., recipes for making lasagna, strategies for fighting wars or (for purposes of this book) methods for managing projects. Whatever their size, the one

Figure 9-2. PM Software Helps Get Your Ducks in a Row

common thread that all projects have is their need to be managed. Without effective planning, progress monitoring, communications and risk management, projects are doomed to fail. The most frequent cause of project failure comes as a consequence of not adopting a proper project management methodology. Project methodologies are systems of practices, techniques, procedures and rules which project managers use to design, plan, implement and achieve their project objectives. Trying to run a project without the aid of a management method is like sailing on the ocean without a compass. Tried-and-tested project management methods help you to plan your project effectively, (steering the work towards the project's goals), while keeping everything ship-shape and headed in the right direction. Different project management methodologies benefit different project types. Adopting the right method ensures that your projects will be planned, controlled, meet specs, be completed on time and stay within the project budget. As you now know, project management (as a profession), hasn't been on the scene for that long, but during its short history its structure, application and methods used for managing it, have grown by leaps and bounds. Many methodologies have been developed and evolved to effectively manage the profusion of different project types and complexities. There is no "one size fits all" method for managing projects, and each approach has its own strengths and weaknesses. A project management methodology should be chosen for its ability to best meet organizational goals and values, address the risks involved, take into consideration the needs of the stakeholders and deliver the best bang for your buck. Although there are many more choices available, four of the most popular today are Agile, critical path method (CPM), PMI/PMBOK (though not everyone agrees that the guide is a methodology), and Waterfall. Here is a little background on each of them:

Agile—Agile is a highly interactive project management methodology, commonly used in software development projects that came on the scene in 2001. Designed for use in projects calling for flexibility and speed, Agile may be best suited for projects requiring less control and real-time communication within self-motivated team settings. Projects using the methodology are not governed by a pre-planned process, but rather are conceived, executed and adapted, as their conditions and situations change. Composed of short delivery cycles (sprints), this allows immediate feedback, provides fast turnaround and reduces complexity. Its four main values are expressed as:

- Individuals and interactions over processes and tools
- Working software over comprehensive documentation
- Customer collaboration over contract negotiation
- Responding to change over following a plan

Critical Path Method (CPM)—The critical path method (developed in the 1950s) is a methodology used for projects with interdependent activities and is based on the concept that there are some tasks you can't start until a previous one has been finished. Stringing dependent tasks together (from start to finish) plots out what is referred to as the critical path. It outlines critical and non-critical activities by calculating the "longest" (on the critical path) and "shortest" (float) time to complete tasks, to determine which activities are critical and which are not. Focusing on a project's critical path allows project managers to prioritize and allocate resources to get the most important work done, and reschedule any lower priority tasks.

PMI/PMBOK—PMI's project management methodology is a set of standards which refers to the five process steps (initiating, planning, executing, controlling, and closing) of project management, which are outlined in the Project Management Body of Knowledge (PMBOK). It provides a framework of conventions, processes, best practices, terminologies, and guidelines that are accepted as standards within the project management industry.

Waterfall—The waterfall methodology (often referred to as the software development life cycle) entails the simple approach of preplanning every step, then executing them, as opposed to the Agile method, which allows changes to customer needs and priorities. Before beginning any work, the project requirements are defined (at the top of the waterfall). Each step is laid out and work is executed in a strict sequence (like water down a waterfall) as the project proceeds through its phases. Without overlap, each phase must be completed before the next phase can begin (the outcome of one phase is the input for the next phase). Any changes are made using a change request. A waterfall approach can provide predictable budgets, timelines and project scopes, and project plans can be easily replicated for future use.

PUTTING TOGETHER A PROJECT MANAGEMENT PLAN

The key to a project's success is its launch from a detailed and well-thought out project management plan (PMP). Such plans are formal documents that spell out how a project is to be implemented, executed, monitored and controlled. As a general rule, PMPs are developed, refined, approved, and frequently revisited and updated (as necessary) to keep the project moving forward, on time and on track. There are templates available (again, some free, some not) that can aid project managers in their creation but, every plan is as individual as the project it lays out and the individuals completing it. They can be simple summaries of the work to be performed or detailed documents, replete with sets of baselines; i.e., scope, schedule and cost (remember the triple constraints?), and/or a number of the subsidiary management plans that form part of the ten knowledge areas found in the PMBOK.

- The Scope Management Plan
- The Schedule Management Plan
- The Cost Management Plan
- The Communication Management Plan
- The Process Improvement Plan
- The Staffing Management Plan
- The Quality Management Plan
- The Risk Management Plan
- The Procurement Management Plan
- The Stakeholder Management Plan

A well-constructed planning document should explain the project's purpose, determine its goals and objectives, identify its approach, expound on its constraints and assumptions, spell out its scope, list its milestones, describe its deliverables, post a schedule, determine dependencies, establish a budget, provide a risk assessment, include a work breakdown structure (WBS), point out the tools and techniques to be used, assign roles and responsibilities, define quality, set standards and procedures and include directories for stakeholders and vendors. Depending on organizational structure, formal approval of the plan is typically granted by the project manager, project sponsor, or the project's resource providers.

PRACTICING BEST PRACTICES

Sooo... how do you become a PM expert? The same way you get to Carnegie Hall—practice, practice, practice—not only by practicing the art and science of the profession, but by using benchmarked (proven) management practices, experienced by project managers in a wide variety of industries and projects. Successful projects are derived from the use of effective project management processes that are referred to as "Best Practices." With the vast array of project types and complexities come a potpourri of models and techniques for managing them, covering the areas of planning, communication, defining scope, objectives and deliverables, and tracking project progress, risk and change management. Below are some of the Best Practices I've employed over the years to bring my projects in on time and within budget. Following these best practices might well improve your chances of achieving your objectives; disregarding them will almost certainly end in your project's failure.

Figure 9-3. Practice, practice, practice!

Hold a project kick-off meeting with everyone having a stake in the project, to acquire an understanding of their expectations, clarify the project's goals and objectives and communicate tasks and assignments to team members.

Define the deliverables by using specific, as opposed to general, terms; have the stakeholders (end users) agree on the accuracy of the spec, and using a WBS (work breakdown structure); decide on the resource and budgetary requirements needed to complete the project.

Use a detailed work definition document to avoid uncertainty and confusion as to who is responsible for what work; establish the level of work they will be tasked to perform; initiate a plan for meeting deadlines and hitting milestones; and bring accountability to the project team.

Manage scope or deliverables changes when assumptions become invalid, to avoid cost overruns and schedule faults; decide if and/or when to accept or reject changes and control how the changes will be incorporated into the project. Consider using a CCB (change control board).

Manage risks by identifying their existence early on in the project; determine their impact on the project, estimate the likelihood of each risk occurring, and create a plan for their avoidance or mitigation.

Manage by walking around to enhance communication by soliciting feedback from the project's participants, making work observations.

Define project management procedures up front, outlining the resources that will be used to manage the project, how the team will manage communication, issues, risk, scope, change and quality. Consider using pre-ordained organizational P&Ps (policies and procedures) or generally accepted project management industry standards.

Manage the work plan by reviewing it on a regular basis to identify completed activities, determining how the schedule and budget are progressing, and determine whether or not the project can be completed within the original cost and duration estimates.

Watch out for scope creep from both minor (small changes that accumulate over time) and major changes (totally new functions or deliverables) being added to the project, which can adversely impact your project, increasing its costs and extending its completion date.

Track and report project progress by comparing actual to planned progress; report variations between actual and planned cost, schedule and scope to stakeholders and upper management, making adjustments as necessary.

Document everything that occurs during the course of the project to prevent recurring problems; furnish data for lessons learned and reviews; and provide historical information for possible litigations arising out of project difficulties.

Hold a wrap-up meeting with all project team members at the end of the project to reflect on the project's flaws, accomplishments and aberrations; review lessons learned and detail a list for optimizing subsequent project endeavors.

Resolve issues as quickly as possible

Chapter 10

Understanding the Framework

COUNTLESS VOLUMES EXIST that deal with every aspect of projects and their management. But there is only one definitive text that gives you the wherewithal to take on most projects. Here is what it is and what it isn't. It is the PMI/PMBOK guide—a research-based product of the at-large project management community, which is led by a volunteer group of highly experienced subject matter experts from around the globe. Content in the guide was developed for project managers, by project managers, and was created to help tailor and select what is needed for managing projects. By PMIs own declaration, the guide represents "good practice for most projects, most of the time," and states that it is "a guide of all the things true about project management" (in a single text). It is not a how-to guide. Rather, it is a set of standard terminologies and guidelines (a body of knowledge) for project management. The guide identifies the knowledge requirements of several facets of the project management profession and details inputs, tools and techniques and outputs available for project managers to utilize; but, it's still up to them to select the right approach for the proper execution of their individual projects. PMIs/Project Management Body of Knowledge guide is touted as the preeminent global standard for project management, or to put it in naval parlance, it is the flagship of the fleet.

EXPLORING THE GUIDE'S ANATOMY

The 5th edition of PMIs project management guide is an internationally recognized standard (ANSI/PMI 99-001-2008 and IEEE 1490-2011) which provides the fundamentals of project management, irrespective of the type of project being undertaken. The PMBOK Guide is process-based, meaning it describes work as being accomplished by the utilization of processes. Project management (as described in the guide) recognizes 47 processes that fall into five basic process groups and 10 knowledge areas that are typical of most projects, most of the time. The first two sections of the guide introduce the reader to the key concepts of the project management field, while the third section summarizes the five process groups. Figure 10-1 shows the interrelationship of the processes and process groups.

The purpose and functions of the five process groups are as follows:

Initiating Process Group

Everything having to do with a project is addressed at this juncture; the initial project scope is defined, the project is aligned with organizational objectives, all stakeholders are identified, the project is formally sanctioned and approved by a sponsor, and management chooses the project manager who will lead it. The project charter and stakeholder register are created at this point.

Planning Process Group

In the PMBOK Guide, PMI defines this group as comprising the majority of the 47 processes (24 to be exact). The planning phase includes the development of the project management plan, establishment of the project scope, the assessment of its requirements, constraints, and assumptions; development of a work schedule, the creation of the work breakdown structure (WBS), determination of a project budget; and planning for quality, communication, human resource, stakeholder, risk, procurement and change management. The (PMP) project management plan, risk register and project schedule are created at this phase.

Executing Process Group

Eight processes are involved in this group, which includes: managing the project team and communications, the achievement

	Initiating	Planning	Executing	Monitoring & Controlling	Closing
Project Integration Management	Develop Project Charter Develop Preliminary Project Scope	Develop Project Management Plan	Direct and Manage Project Execution	Monitor and Control Work Integrated Change Control	Close Project
Project Scope Management		Scope Planning Scope Definition Create WBS		Scope Verification Scope Control	
Project Time Management		Activity Definition Activity Sequencing Activity Resource Estimating Activity Duration Estimating Schedule Development		Schedule Control	
Project Cost Management		Cost Estimating Cost Budgeting		Cost Control	
Project Quality Management		Quality Planning	Perform Quality Assurance	Perform Quality Control	
Project HR Management		Human Resource Planning	Acquire Project Team Develop Project Team	Manage Project Team	
Project Communications Management		Communications Planning	Information Distribution	Performance Reporting Manage Stakeholders	
Project Risk Management		Risk Management Planning Risk Identification Qualitative Risk Analysis Quantitative Risk Analysis Risk Response Planning		Risk Monitoring and Control	
Project Procurement Management		Plan Procurement Management	Conduct Procurements	Control Procurements	Close Procurem ents
Project Stakeholder Management		Plan Stakeholder Management	Manage Stakeholder Engagement	Control Stakeholder Engagement	

Figure 10-1. Processes and Process Groups Matrix

of the project's objectives, procuring and managing resources, executing defined tasks, implementing approved changes, ensuring quality assurance and managing stakeholder engagement. Most of the deliverables will be produced in this phase, and the bulk of the budget will be expended during the project execution phase.

Monitoring and Controlling Process Group

There are 11 processes that comprise the monitoring and controlling process group. The PMBOK Guide, states that these are the "processes required to track, review and regulate the progress and performance of the project; identify any areas in which changes to the plan are required; and initiate the corresponding changes." They include the control of (and managing changes to) a project's scope, quality, risks, cost, schedule, communications, procurements and stakeholder engagement, as well as measuring project performance, utilizing appropriate tools and techniques, and ensuring conformance to standards. This process group is where the risk registers and risk response plan are updated, corrective actions on the issues register are assessed, and project status is communicated to the stakeholders.

Closing Process Group

The fifth group only involves two processes: closing procurements and closing the project (or phase) which involves final completion of all project activities, including the acceptance of deliverables, archiving of documents, and communication of project closure. Other activities include the attainment of financial, legal, and administrative closure; closing out contracts, making final payments, discussing learned lessons and distribution of the final project report.

Sections 4 through 13, comprising the main core of the book (the heart of PMBOK and project management), explain how to manage and execute a project in each of the process groups, focusing on each of the project management knowledge areas. You may recall that I provided a cursory definition of each knowledge area in Chapter 3, but it would behoove you to review the 347 pages devoted to them in the PMBOK, to gain a more thorough understanding of the breadth the 10 areas covered. Each of the ten knowledge areas contains the processes that need to be accomplished within its discipline to achieve effective project management. Each of these processes also falls into one of the five process groups, creating a matrix structure such that every process can be related to one knowledge area and one process group. (See Figure 10-1.)

THE EVOLUTION OF THE GUIDE

The PMBOK (Project Management Body of Knowledge) guide outlines a set of standard terminologies and guidelines for the profession of project management. It was first published in 1996, followed by successive editions in 2000 (2nd Edition), 2004 (3rd Edition), 2009 (4th Edition), 2013 (5th Edition) and most recently the 6th edition (at this writing) in September of 2017. Each subsequent release was written to surpass the relevant content of its prior version, with the application of new standards and best practices as they became manifest in the project management field. The 5th edition of the guide is considered an important exam preparation text for PMIs—PMP (Project Management Professional) certification. **Note:** In September of 2017, the PMBOK 6th edition launched, but will not be utilized for exam preparation until later in 2018.

In its infancy, PMI determined that there was a need to combine existing project management documents and guides to bolster the development of the project management educational process. Expanding on a version of a 1983 white paper titled "Ethics, Standards, and Accreditation Committee Final Report," they published the first edition of the PMBOK® Guide in 1996. When the guide was upgraded in the year 2000, the 2nd edition included proven project management knowledge and practices that were commonly accepted in the field and found to be reliable and useful for most projects. The added content emulated the growth of the profession, and the guide was improved through the removal of errors found in the prior edition. In the 3rd version, as a consequence of PMI's editorial committee review of thousands of suggestions for improvements to the guide, they integrated many of the recommendations and published an upgrade in 2004. A more consistent and accessible 4th edition was published five years after that, where a clear distinction was made between the project documents and the project management plan. The 5th edition, released in 2013, represents PMI's latest update to the project management body of knowledge, just short of the 6th edition upgrade. The PMBOK® Guide 6th edition (published in September of 2017) incorporates Agile (one of the fastest growing methodologies) in its module, as well as the PMI Talent Triangle (Leadership, Technical Project Management, Business and Strategic Management) and includes some minor changes in the process groups and processes.

Note: Earlier versions of the PMBOK Guide were recognized as standards by the American National Standards Institute (ANSI) which assigns standards in the United States (ANSI/PMI 99-001-2008) and the Institute of Electrical and Electronics Engineers (IEEE 1490-2011).

Author's note: Although the PMBOK Guide is not a step-by-step manual for implementing projects, the information it contains can augment a project manager's understanding of the entire accumulated project management body of knowledge, enabling them to choose and use an appropriate methodology.

The history of the Project Management Institute (PMI®) began with its inception at its first meeting held in Atlanta, Georgia, in 1969. The first credential issued by PMI was the Project Management Professional (PMP), awarded in 1984 with the administration of the first PMP® exams. In 1987, the "Project Management Body of Knowledge" was released (sections A through H, 5-6 pages in each section). In August of 1994 a 64-page exposure draft of the PMBOK® Guide was released that contained eight knowledge areas. The exam consisted of 320 questions (40 from each of the eight knowledge areas), and the test taker was allowed 6 hours to complete the exam. The first edition (1996) of the PMBOK® Guide had 176 pages, detailing nine knowledge areas (integration management was included for the first time) and 37 processes. Henceforth, every fourth year, a new edition was published (see Figure 10-2). In the year 2000, the 211-page, second edition was released with nine knowledge areas and 39 processes. In December 2004, the 390-page third edition was released with five new processes (bringing the total to 44) and 592 inputs, techniques and outputs (ITTO) were added. The 467-page 4th edition, released four years later (December 2008) had 2 less processes (42) and 75 less ITTO (517). In December 2012, the current 589-page 5th edition was released with 47 processes and 619 ITTO.

Figure 10-2. At the Crossroads (Every 4 Years—or So)

With each release of a new edition, PMI revises its examination for the PMP certification, to ensure that it accurately reflects the knowledge and skills needed by project management professionals for leading present-day projects. The 6th edition was released in September of 2017, and the PMP exam will change 26 March 2018. As with prior iterations, the 6th edition has been updated to reflect the latest good practices in project management and to ensure the exam content is consistent with the PMBOK® Guide. To each knowledge area, a section has been added called "Approaches for Agile, Iterative and Adaptive Environments," which describes how these practices integrate in project settings, contains more emphasis on strategic and business knowledge, includes discussion of project management business documents, and provides information on the PMI Talent Triangle™. Major changes in the PMBOK Guide 6th edition are the inclusion of Agile Project Management knowledge and practices, a practice guide published on Agile project management and re-arrangement of the three introductory sections.

EXTENSIONS TO THE PMBOK GUIDE

The PMBOK Guide is a general embodiment of information useful to project managers for managing most projects, most of the time. Augmenting that data, PMI currently offers three official extensions covering the industry specific focuses of software, construction and government project management. A fourth extension (U.S. Department of Defense Extension to: A Guide to the Project Management Body of Knowledge [PMBOK® Guide]) was released in June of 2003, but is currently not available as a PMI standard on its site. Following is a summarization of each of them.

Software Extension to the PMBOK® Guide Fifth Edition

Library of Congress Cataloging-in-Publication Data
ISBN-13: 978-1-62825-013-8
ISBN-10: 1-62825-013-5

Taking the complexity of present-day software development into consideration, the PMBOK® Guide editions lack the detail and specificity required to address the unique context and needs of software development projects. To counteract the problem, PMI partnered with the IEEE Computer Society to publish the Software Extension to the PMBOK® Guide. The software extension (5th edition) is a detailed, comprehensive guide that covers the full spectrum of project approaches applicable to software development. One of the most important titles PMI has ever produced, it provides a solid foundation for understanding predictive and adaptive project life cycles. Written in the style (and following the structure) of the PMBOK Guide, the book is useful to project managers, system analysts, and software architects in pursuit of software development, providing project managers with the knowledge and practices they need to improve their efficiency and effectiveness, and that of their project team members. The work describes processes that are applicable for managing adaptive lifecycle software projects and includes coverage of project, scope, time, cost, quality, communications, risk, and procurement management. It is designed to be used in tandem with the latest edition of the PMBOK® Guide, providing readers with a balanced view of methods, tools, and techniques for managing software projects across the life cycle continuum. The software extension uses the same section and subsection organization found in the PMBOK® Guide. There are sections for project life cycle, project management processes, project integration, project scope, time, cost, quality, human resources, communications, risk, procurement, and stakeholder relations. While this extension focuses on the management of software development projects, it can also be useful for managing IT projects. Such projects may require in-house development of application software for which this extension relates directly, as most of the organizational and team materials in the extension apply equally to information technology development.

Construction Extension to the PMBOK® Guide Third Edition

Library of Congress Cataloging-in-Publication Data has been applied for.

ISBN: 978-1-62825-090-9

The Construction Extension is a useful reference tool for construction project management practitioners, students, and interested professionals, which adds to the general project management knowledge and practice on construction sites that are not otherwise covered in the PMBOK® Guide, and eliminates specific processes and references that become obsolete with each new edition. It first appeared in 2003 and has been updated every time the PMBOK® Guide has. As is the case with the other extensions, the construction extension is designed to be used in tandem with the latest edition of the PMBOK® Guide. The current edition follows the structure of its previous editions and keeps the chapter structure of the parent guide, with preface, introductory chapters, knowledge areas, annexes, and appendixes. The organization of the knowledge areas follows the process groups' scheme of initiating, planning, executing, monitoring and controlling, and closing. The first chapter establishes the framework for the remainder of the document. Chapter 2 homes in on the context, addressing specific topics such as types of projects, delivery methods, and organizational environmental factors. Chapter 3 begins by mirroring the same type of content as its equivalent in the PMBOK® Guide, giving a brief introduction to the content of each knowledge area before presenting some important trends in the industry and advances in construction project management. This section features topics such as building information modeling (BIM), integrated project delivery (IPD), global markets, and ethics. In the knowledge areas, integration management includes a topic on construction administration; cost management covers escalation, inflation, and currency exchange; scope management argues the need to establish clear limits; schedule management highlights the linear scheduling method; resource management addresses human resources, machinery, tools, equipment, and materials; communications management explores document control issues; risk management talks about insurance; quality management speaks to non-conformance; procurement management emphasizes the hierarchies of buyers and sellers; and stakeholder management looks at the prominence of labor unions and insurance providers. The back matter lists PMI's references and a glossary of construction terms and acronyms (complementary to the ones featured in the PMBOK® Guide) that are used in the extension.

Government Extension to the PMBOK® Guide Third Edition

Library of Congress Cataloging-in-Publication Data
ISBN 13: 978-1-930699-91-5 ISBN 10: 1-930699-91-3
The Government Extension provides a framework for ensur-

ing effective and efficient management of projects in the public sector. Derived from the PMBOK Guide, it is specifically designed to suit the unique characteristics of government-relevant projects, by interpreting and extending the precepts of proficient project management found there. The extension in the PMBOK Guide contains valuable and useful attributes (recognized as good practice) that are germane to national and state-owned organizations. As is the case with all of PMI's guides, the information they contain is there to enlighten project managers who are ultimately responsible for determining its application in their projects. To effectively manage them, the project management team must recognize the legal constraints (projects may be subject to certain laws and regulations) on government projects, their accountability to the public (project managers are accountable to many stakeholders beyond the immediate client), and the utilization of public resources (government budgets are funded with public resources). The extension provides guidance for managing the intricacies of government projects, focuses on principles for ensuring effective project controls and enables the accountability of project expenditures to a nation's citizens, as required by public law.

U.S. Department of Defense (DOD)
Extension to the PMBOK® Guide

First Edition; Version 1.0, June 2003
Published by the Defense Acquisition University (DAU) Press; Fort Belvoir, Virginia 22060-5565

This U.S. Department of Defense (DoD) Extension was developed, is published, and will be maintained under agreement between the Project Management Institute, Inc. (PMI®) and the Defense Acquisition University (DAU) as the U.S. DoD Extension to PMI's "A Guide to the Project Management Body of Knowledge" (PMBOK® Guide). The extension has been approved as a PMI Standard™ through the PMI Standard-Setting Process. The current document is the first edition (2003) of the U.S. Department of Defense (DoD) Extension which I have personally used for guidance when managing defense-related projects, since the issuance of my PMP (Project Management Professional) certification in that year. Its primary purpose is to identify and describe defense applications of the core project management knowledge areas contained in the PMBOK® Guide, as well as those defense-intensive knowledge areas not contained in the Guide, and is not intended to be a stand-alone document. To the extent that the U.S. DoD Extension is a derivative work from the PMBOK® Guide, PMI has granted DAU a limited, non-exclusive, non-transferable royalty-free license (the "License") to prepare and publish the U.S. DoD Extension as a derivative work based upon the PMBOK® Guide. The U.S. DoD Extension is a "work of the United States Government" under the provisions of 17 U.S.C. §105. However, while PMI does not object to publication of the U.S. DoD Extension in "the public domain," such consent in no way waives PMI's underlying copyrights to the PMBOK® Guide. The work to develop the first edition of the U.S. DoD PMBOK Extension was funded by DAU. At the time of its release, it was agreed (for purposes of updating the U.S. DoD Extension as a PMI Standard) that when significant additions or other changes to the content of the U.S. DoD Extension were made, DAU would notify PMI of those changes and submit the revised U.S. DoD Extension for review, pursuant to the PMI Standards Setting Policy and Procedures. The document was originally written to complement and supplement the 2d edition of the PMBOK Guide. Although the first edition (of the extension) has been well received, especially from the international program management and acquisition community, it was never used as a textbook by either PMI or DAU—as had been envisioned when the document was approved for development. Accordingly, when funding for DAU was cut in 2005 and 2006, the Extension (and its update) became a casualty of those cuts. DAU has no plans for updating the U.S. DoD Extension but, it is rumored that the Aerospace and Defense Specific Interest Group of PMI is considering a volunteer effort to update the existing extension.

DERIVING THE DOCUMENTS

There are a great number of documents created and utilized for managing project work. Many of them are considered as helpful tools for organizing a project's structure and course, while others are determined to be absolutely indispensable for their successful completion. Both types are addressed in the PMBOK (Project Management Body of Knowledge) and are listed below (alphabetically) with a brief summary and a reference to where they can be found in the Knowledge area of the text.

Author's note: Bear in mind that not all documents are needed for all projects. Over utilization can be as counter-productive as underutilization of these tools.

Activity Cost Estimates—Quantitative assessments of the probable costs required to complete project work. They include labor, materials, equipment, fixed cost items like contractors, services, facilities, financing costs, etc. (Project Cost Management) 7.2.3.1

Activity Duration Estimates—Useful in identifying risks related to the time allowances for the activities or project. In project scheduling, each activity is assigned a duration after the resources are determined. (Project Time Management) 6.5.3.1

Activity List—Quantitative assessments of the likely number of time periods that are required to complete an activity and the official breakdown of the project activities into tasks. (Project Time Management) 6.2.3.1

Activity Resources—Identify the types and quantities of resources required for each activity in a work package. (Project Time Management) 6.4.3.1

Agreements—Applicable agreement information and costs relating to products, services, or results that have been or will be purchased are included when determining the budget. (Project Procurement Management) 12.2.3.2

Basis of Estimates—The amount and type of additional details supporting the cost estimate vary by application area containing things like assumptions, constraints, range (plus or minus a percentage), confidence level, etc. (Project Cost Management) 7.2.3.2

Change Log—Used to document changes that occur during a project and their impact to the project in terms of time, cost and risk. (Project Integration Management) 4.5.3.2

Change Requests—A formal proposal to modify any document, deliverable, or baseline by a stakeholder wanting the project manager to make changes to the project. (Project Integration Management) 4.3.3.3

Communications Management Plan—Describes how project communications will be planned, structured, monitored, and controlled. (Project Communications Management) 10.1.3.1

Cost Management Plan—Describes how the project costs will be planned, structured, and controlled and includes items such as level of accuracy, control thresholds, and performance measurement. (Project Cost Management) 7.1.3.1

Human Resource Management Plan—This document describes how the project team will be defined, acquired, managed and eventually released and includes information such as organizational charts, roles and responsibilities, resource calendars, and management techniques. (Project Human Resource Management) 9.1.3.1

Issue Log—Used to document and monitor the resolution of issues, to facilitate communication and ensure a common understanding of issues. (Project Stakeholder Management) 13.3.3.1

Make-or-Buy Decisions—A make-or-buy analysis results in a decision of whether particular work can best be accomplished by the project team or needs to be purchased from outside sources. (Project Procurement Management) 12.1.3.5

Process Improvement Plan—Details the steps for analyzing project management and product development processes to identify activities that enhance their value. (Project Quality Management) 8.1.3.2

Procurement Documents—Documents that solicit bids, quotations, or proposals from interested vendors (aka, Invitation to Tender, Request for Proposal (RFP), Request for Qualifications (RFQ), etc. (Project Procurement Management) 12.1.3.3

Procurement Management Plan—Describes what external expertise will be required, its justification and scope. (Project Procurement Management) 12.1.3.1

Procurement Statement of Work—Describes the work to be performed under a contract. Different industries often have variations which have specific meanings, like Terms of Reference, etc. (Project Procurement Management) 12.1.3.2

Project Charter—Initiates the project, authorizes the project manager and documents the business need for the project as well as assumptions and constraints the project must live under. (Project Integration Management) 4.1.3.1

Project Communications—Involves the activities that are required for information to be created, distributed, received, acknowledged, and understood, including but not limited to performance reports, deliverables status, schedule progress, and cost incurred. (Project Communications Management) 10.2.3.1

Project Funding Requirements—Includes projected expenditures plus anticipated liabilities by taking the Activity Cost Estimates and adding the additional factor of time. (Project Cost Management) 7.3.3.2

Project Management Plan—Outlines how the project will be managed, and includes the project schedule, budget, quality standards, project team requirements, project control, and anything else that is necessary to communicate how the project will be managed. (Project Integration Management) 4.2.3.1

Project Schedule—Contains information on required timelines or mandated deliverable dates; usually displayed as a Gantt chart. (Project Time Management) 6.6.3.2

Project Staff Assignments—The staffing documents for the project team, including staff directories, organizational charts, etc. (Project Human Resource Management) 9.2.3.1

Quality Checklists—Used to conduct the inspections to confirm the acceptance or rejection of the products based on the predetermined quality metrics. (Project Quality Management) 8.1.3.4

Quality Control Measurements—Measurements of the quality of deliverables produced by the project. (Project Quality Management) 8.3.3.1

Quality Management Plan—Describes how the organization's quality policies will be implemented, defines the quality standards against which the products will be measured, how they will be measured, and lists the pass/fail criteria. (Project Quality Management) 8.1.3.1

Quality Metrics—describes a project or product attribute and how the control quality process will measure it and outlines the project or product attributes which will be monitored and controlled. Project Quality Management) 8.1.3.3

Requirements Management Plan—Describes how requirements will be analyzed, documented, and managed. (Project Scope Management) 5.1.3.2

Requirements Traceability Matrix—Tracks each requirement in the requirements management plan to ensure all of the small details are addressed and the requirements are satisfied. (Project Scope Management) 5.2.3.2

Resource Breakdown Structure—A description of resources (such as labor, material, equipment, and supplies), by category and type. (Project Time Management) 6.4.3.2

Resource Calendars—Contains information on the availability of staff members and resources during the project. (Project Human Resource Management) 9.2.3.2

Risk Management Plan—Describes how risk to the project's success factors will be managed and includes things like risk analysis methodologies, budgeting, and definitions of risk probability and impact. (Project Risk Management) 11.1.3.1

Risk Register—Contains a list of the most important risks to the project's completion and identifies the likelihood of occurrence, the impact to the project, the priority, and response plans where applicable, and details the initial response to the risk; i.e., acceptance, avoidance, mitigation, or transfer. (Project Risk Management) 11.2.3.1

Schedule Data Metadata—Includes information such as technical data, potential changes, and information for the project manager. (Project Time Management) 6.6.3.3

Schedule Management Plan—The central planning document relating to project scheduling, containing things like the scheduling methodology and tools, as well as level of accuracy, units of measure, and organizational procedures. (Project Time Management) 6.1.3.1

Schedule Network Diagram—A graphical representation of the logical relationships between the tasks within the project. (Project Time Management) 6.3.3.1

Scope Management Plan—Outlines how the project scope will be managed, how scope changes will be addressed, and how the project scope will be monitored and controlled to ensure scope changes do not happen unless they are required. (Project Scope Management) 5.1.3.1

Scope Statement—Outlines the scope of the project the primary project deliverables as well as expressly includes or excludes items that are not obvious to the scope. (Project Scope Management) 5.3.3.1

Source Selection Criteria—Includes information on the supplier's required capabilities, capacity, delivery dates, product cost, life-cycle cost, technical expertise, and the approach to the contract and identifies the criteria under which the bids, quotations or proposals will be evaluated. (Project Procurement Management) 12.1.3.4

Stakeholder Analysis—Analyzes each project stakeholder in the categories of Power (to create project changes) and Interest in the project. (Project Stakeholder Management) 13.1.2.1

Stakeholder Management Plan—Describes how the stakeholders will be managed and includes things like the stakeholder engagement levels, communication requirements, and stakeholder analysis. (Project Stakeholder Management) 13.2.3.1

Stakeholder Register—Identifies each project stakeholder, including which part of the project scope is of the most interest to them, what their main requirements are, their expectations and potential influence. (Project Stakeholder Management) 13.1.3.1

Team Performance Assessments—Formal or informal documentation of the project team's performance to resolve issues, modify communication, address conflict, and improve team interaction. (Project Human Resource Management) 9.3.3.1

Work Breakdown Structure—The breakdown of a project into components for the purpose of identifying the scope. (Project Scope Management) 5.4.3.1

Work Performance Reports—Provide documentation about the current project status compared to project forecasts. (Project Integration Management) 4.4.3.2

MANAGING PROJECTS WITH THE PMBOK GUIDE

Without formal training in the profession, most folks are ill equipped to manage the projects they are assigned, most often relying solely on the management skills they possess to get them through. There are untold amounts of project management literature available to them, but how can they acquire the knowledge that they need to get the job done—especially when tasked to perform the work under unfavorable time constraints? The fact is there is no single volume on project management that spells out everything there is to know about the subject; but, if push comes to shove, PMI's Project Management Body of Knowledge is as close as you'll get to such a compendium of information. Through the iterations of the publication, its logical, structured and reasoned approach to the processes involved in managing projects has evolved and continues to advance the practice. The PMBOK guide is crammed with relevant information needed by project practitioners (with a modicum of training) to bring simple projects in, on time and under budget. I used the guide as my assist, well before I studied for and acquired my PMP certification.

Following the phasing outlined in the guide, I read through the text and compiled my overarching strategy for undertaking and completing project work. In the initiation phase, I developed a simplified project charter, derived from a template I found online, to establish a business case for a project that defined why it was being undertaken and what problem it was meant to solve. In the planning phase, I enlisted all of the knowledge areas while developing a project management plan (again utilizing web based materials to put the document together, laying out how the project would be managed, executed and controlled. In it I included a scope statement to define the project boundaries, a work breakdown structure (WBS), a project schedule and budget, and plans derived from the knowledge areas in the PMBOK. For the execution phase, I outlined how I would manage the project team, determine my communications strategy, conduct procurements and manage quality and stakeholders. The items I listed for the controlling phase included costs, scheduling, scope, quality and procurement. Finally, I took into account project documentation, closure of contracts with consultants, contractors and suppliers, the equipment and systems commissioning process and reporting in the closing phase.

Chapter 11

Experience is the Best Teacher

Early on in my career, I took on the challenge of project management, first as an ad-hoc (incidental) PM, then becoming more polished as I took on bigger and more involved ventures. Every project I completed was a stepping stone to the next, where I applied the knowledge I acquired and the lessons I learned from my prior undertakings. In the prior 10 chapters, I talked about several projects that I undertook in my career, briefly outlining some of the concerns I had, sharing a cursory understanding of my early approaches to managing them and citing things that I learned for and from doing.

In my description of the parking lot project, I spoke to the fact that there is much more to managing projects than what might first be assumed, the need for learning the geometrics involved, acquiring an appreciation for the impact on ancillary departments, and the value of utilizing existing managerial skills sets.

When I discussed the emergency treatment suite project, I talked about incorporating drawing software as a management tool, my first attempt at design, and my need for gaining a better understanding of the various utility systems, their components and interconnections affecting the project's outcome.

During the family practice residency project, I learned a great deal about materials, the relevance of functional adjacencies between different departments and their operational inter-relationships. Each time I took on a new assignment, I analyzed all aspects of the proposed work, applied the lessons I learned from my past endeavors and prepared myself (as best as I could) for completing it by researching pertinent regulatory requirements (germane to the project), obtaining the required subject matter knowledge and determining which principles and practices to apply. As the projects became more sophisticated and complex, it was incumbent upon me to upgrade my knowledge base, become more familiar with codes, standards and practices, and more proficient with project management tools and techniques. The following are (summary) breakdowns of some of the larger, more diverse and advanced projects I've been involved with over the years. Each of them will state the scope of the project, associated duties, the documents I used, how I prepared for them and what I learned from the experience. They cover the categories of equipment installation, software implementation, organizational program management, commercial construction and government contract work.

INSTALLATION OF EQUIPMENT

Spending 4 years as a U.S. Navy boiler technician gave me the engineering foundation I needed to pursue career aspirations in the real world, setting the stage for my acquisition of a Chief Engineers (operator) license and a Certified Plant Engineer (CPE) certification. Working in the electro-mechanical field prepared me (in large part) to lead projects involving the selection and installation of various types of equipment, as a facility manager. Some of the equipment installs included:

Project(s) Title:

Installation of Diagnostic and Emergency Power Equipment

Summary Description:

Construction of concrete pads and installation of 480V service entrances (Crouse-Hinds) electrical umbilical (with weatherproof entryways)

Staging of trailer mounted CT (computer tomography) and magnetic resonance imaging (MRI), scanners and lithotripters

Construction of a concrete block enclosure for housing emergency power equipment

Placement of a diesel-powered generator

Installation of all associated electrical switchgear

Connection of the generator with normal house power circuitry

My Position:

Project Manager

Associated Duties:

Solicited AEC (architect, engineering and construction) firms

Prepared RFPs (requests for proposals) for the purchase of equipment

Acquired and reviewed blueprints

Coordinated installations with appropriate department heads

Oversaw construction activity

Preparation:

Explored contract types
Researched applicable codes
Studied equipment primers
Met with contractors and department heads

Documentation:

Activity cost estimates
Activity resources
Cost management plan
Issue log
Procurement documents
Project charter

Quality checklists
Schedule management plan
Scope statement

Learning Opportunities:

Study of high-voltage electricity

Study of diesel engines

Interactions with AEC personnel

Study of gauss units and fiberglass rebar (reinforcement rods for insertion in concrete) for avoidance of magnetic field exposure problems

SOFTWARE IMPLEMENTATION PROGRAMS

Again, over the course of my many years as facility manager, utilizing my equipment operations background as a basis of understanding, I combined my knowledge of equipment with the study of computerized maintenance management systems (CMMS), to install and implement software for overseeing the care and repair of building equipment and systems at different facilities. A CMMS, sometimes referred to as enterprise asset management (EAM), is computer software designed to simplify the management of building systems support maintenance. The application had its beginnings in the late 1970s; as organizations began to migrate from pencil/paper management to more dependable tracking by computers, then became prevalent in the late 80s and early 90s.

Project Title(s):

Preventative Maintenance Program(s)

Summary Description:

Installation and implementation of multiple iterations of Infor, Maximo, Panda Plus and Hems CMMS (computerized maintenance management system) software programs, for the establishment of preventive maintenance practices at several different facilities

My Position(s):

Participant/Project Manager

Associated Duties:

Data collection from building equipment

Transfer of information from written records

Establishment of policies and procedures

Preparation:

Researched costs and features of available software

Queried companies regarding software maintenance policies

Included operator training in the purchase price

Put together an in-house conversion plan

Documentation:

Activity resources

Human resource management plan

Procurement document

Project communications

Project management plan

Project schedule

Work breakdown structure

Learning Opportunities:

Trained on the abilities and use of CMMS software

Increased my understanding of computer operations

Updated my knowledge of preventative maintenance

ORGANIZATIONAL FACILITY AND PROGRAM MANAGEMENT

As project overseer in many organizations, I've been responsible for multi-million dollar operating and construction budgets, solicitation of bids and awarding contracts, directing owners' meetings, conducting audits, establishing contingency funds, performing reserve studies, ruling on change orders, concluding punch lists, completing comprehensive plant surveys and post occupancy evaluations (POE) inspections, writing specifications, negotiating myriad service agreements, as well as developing and supervising in-house construction crews and projects. Many of the actions I've taken involved the assessment and reorganization of facility management departments, resulting in the creation of policies, manuals, and corrective action plans. Here are some examples of that work:

Project Title:

1. Internet Service Exchange (ISX) Build Outs—[DATA CENTERS]
2. Facility Organization Infrastructure

Summary Description:

1. Complete the gutting of six existing structures (closed department stores and supermarket) and Greenfield construction of comprehensive physical plant infrastructures for housing internet service providers. **Author's note:** A Greenfield project is one that lacks constraints imposed by prior work. The analogy is to that of construction on Greenfield land where there is no need to work within the constraints of existing buildings or infrastructures.
2. Create an organizational infrastructure program for each of the facilities.

My Position(s):

East Coast Regional Facility Director/Construction Coordinator

Associated Duties: (construction related)

Coordinated all aspects of the facilities infrastructure layout and utilities with the construction arm of the company

Considered facility, environmental, safety and health considerations in all of the East Coast ISX's

Reviewed and commented on all MEP (mechanical, electrical, plumbing) drawings

Attended and inputted at owner's meetings

Witnessed testing of centrifugal chillers (6,000-ton total capacity) and twenty-two 1.67 MW generators

Negotiated with power companies for the installation of secondary electrical feeds

Performed value engineering of specified systems (rotary UPS and humidifiers)

Reviewed NDE (non-destructive examination), welding qualifications (Section IX—ASME) and ASME (code) documents

Conducted regular inspections (walk-throughs) with the electrical and mechanical contractors

Associated Duties: (managerial)

Authored equipment and systems service specifications (water treatment, etc.)

Authored an RFP (request for proposal) for acquisition of facility staffing

Reviewed, approved and witnessed a five-level system commissioning process

Wrote standard operations procedures (SOP) manuals

Generated equipment operating instructions

Created job descriptions and employee work schedules for

all facility management department positions

Authored a department safety manual

Preparation:

Separate kickoff meetings with the construction arm of the organization and the *West* Coast Regional Director

Attendance at weekly construction progress meetings

Frequent visits to the intended worksites

Preliminary coordination with the electrical utility companies

Interviewed potential facility manager candidates for oversight of the data centers

Documentation:

Activity lists
Agreements
Change log
Communication management plan
Human resource management plan
Issue log
Project management plan
Quality checklists
Schedule management plan
Team performance assessments
Work breakdown structure

Learning opportunities:

Observational visits to existing data centers

Conferences with data center facility managers

Familiarization with the national BOCA (Building Officials Code Administrators) and SBCCI (Southern Building Codes Congress, Inc.) building codes

Familiarization with AIA (American Institute of Architecture) standards

Learned the 16 divisions of the Uniform Construction Code

Worked with NFPA 101 (Life Safety Code), NFPA 70 (National Electrical Code) and ADA requirements

Gained an understanding of the "grandfather clause" waiver process

COMMERCIAL CONSTRUCTION PROJECT(S)

Throughout my career, my construction background spanned both the commercial and government sectors. I'll cover two commercial projects here before delving into the government contracting arena in the last section of this chapter. Earlier in the book, I touched on a $7 million project I oversaw involving the construction of a women's (OB/GYN) pavilion and an in-house laboratory expansion. As administrative director, I was the hospital's liaison between administration, county government and the (AE) architectural and engineering firms and general contractor for both projects. The following is a more in-depth breakdown of that (combined) project.

Project Title:
1. Women's (OB/GYN) Pavilion
2. In-house Laboratory Expansion

My Position:

Project Manager

Summary Description:
1. Construction of a 30,000-square-foot women's medical services wing
2. Expansion of the existing medical laboratory

Associated Duties:

Reviewed and approved all drawings (partials/as-builts)

Determined that all required permits were acquired and properly posted

Conducted daily walk-throughs with the building superintendent and safety officer

Conducted weekly owner's meetings

Wrote and distributed monthly progress reports

Conferred with the cost estimator, scheduler and end users

Reviewed and approved petitions for modifications and change orders

Approved allocations of contingency funds (errors and omissions)

Coordinated 24/7 department operations for critical system tie-ins

Arranged for contingency waivers for circumventing "Life Safety Code" requirements

Accompanied inspectors on their rounds

Assembled the final punch list items

Assured that policy was in place and regulatory compliance requirements were strictly adhered to

Determined the point of "substantial completion"

Preparation:

Solicited bid requests
Selected the A&E firms
Approved final contract award
Oversaw construction and approved project completion
Developed an in-house (high-end) construction capability

Documentation:

Agreements
Change, issue and risk logs

Combined work schedules
Procurement documents
Project charter
Project management plan
Quality management plan
Resource calendar
Secondary plans
Statements of work
Stakeholder analysis
Team performance assessments
Work performance reports
Work breakdown structure

Learning Opportunities:

Familiarization with materials submittals

Study of RFIs (requests for information) intent and procedure

Study of various types of construction contracts and the use of subcontractors

Taking of detail and progress pictures

University courses on construction supervision

Project Title:

Medical Center Restoration

Summary Description:

Total gutting and restoration of a 230-bed hospital structure

My Position(s):

Interim vice-president/administrative liaison

Associated Duties:

Provided technical direction during the course of the project, through the licensing, re-opening and accreditation of the institution

Attended all construction meetings on behalf of the stakeholders

Exercised absolute authority governing all electrical, mechanical and/or environmental issues

Reviewed and approved all substitutions, modifications and change orders

Submitted regular progress reports to the county commissioners and hospital administration

Witnessed all testing, including concrete slump tests, hydrostatic pressure tests, air balancing, potable water system, effluent discharge, system commissioning and air emissions

Preparation:

Relied heavily on my past work and management experiences

Met with all stakeholders of the project (including county commissioners, hospital administration and trade union reps)

Researched applicable regulatory requirements

Solicited bid requests

Selected the AEC (architectural, engineering and construction) firms

Approved final contract award

Oversaw construction and approved completion of the project

Documentation:

Agreements
Change requests
Cost management plan
Project management plan (theirs)
Quality assurance plan (theirs)

Risk management plan
Requirements management plan
Stakeholder management plan
Submittal register

Learning Opportunities:

Insight into county government and trade union processes, politics and procedures

GOVERNMENT CONTRACT WORK

All the prior work (and many more projects that I left unmentioned) paid dividends when I finally entered into the government contract arena. Here I became familiar with the government bidding process, and learned to work under IDIQ (indefinite delivery-indefinite quantity), cost plus fixed fee and fixed cost plus award fee contract delivery arrangements. Having worked with POC's (points of contact) from the DoD (Department of Defense) and NASA (National Aeronautics and Space Administration), I also became familiar with the Federal Register; the CFR (Code of Federal Regulations); and the FAR (Federal Acquisitions Regulations). The three government based projects I've chosen to share are: a (second phase of an overall $83 million project), plant improvement of a federal security facility and a naval air station recapitulation program.

Project Title(s):

1. Phase two (2) of an $83 million Federal Institute power plant expansion
2. Construction of a free-standing electrical power substation building

Summary Description:

1. Close out of Phase I of the power plant expansion which included the addition of five 5,000-ton centrifugal chillers,

supporting infrastructure and tie in with the existing plant. Construction management for the building to house two additional chillers, installation of their support equipment, system commissioning and tie-ins to the existing plant.

2. Construction of a separate electrical substation to augment the site's power requirements

My Position:

Senior project manager

Associated Duties:

Administered a construction management team consisting of a project manager, cost estimator, scheduler, contract specialist, inspector and clerical personnel

Acted as liaison between the construction company and the government project officer

Conducted bi-weekly project review meetings

Wrote and distributed progress reports

Performed quality assurance audits and site inspections

Prescribed the parameters to be utilized in the commissioning process

Attended witness-testing activities on completed installations

Completed re-negotiation activities for reconciliation of budgets

Reviewed and approved (or denied) contractor pay requests, change order requests and submittals

Preparation:

Researched the Federal Register, the CFR (Code of Federal Regulations) and the FAR (Federal Acquisitions Regulations) sections, relevant to both project pieces

Documentation:

Utilized the vast majority of documents available

Learning Opportunities:

Masters level university courses in applied project management

Extended studies of government contracting and construction management

Project Title:

Naval Air Station Recapitulation Plan

Summary Description:

Analysis of hangar and runway pavement capacities and condition

Suggested possible relocation of operations tower

Recapitalization plan for hangars, including assessment of existing building support systems and utilities

Capital improvements plan, including flight line ground electronics systems

My Position(s):

Subject matter expert/advisor

Associated Duties:

As a sub-contractor: provided modification and replacement opinions on facility structures, pavements, electrical, mechanical and fire protection systems and utilities service

Determined environmental impacts

Assisted with preparation and submission of the required DoD (Department of Defense) documents for acquiring MilCon (military construction) funding

Preparation:

Met with the contractor and formulated a plan for initiating the project

Attended charrettes with military and civilian base personnel to determine the project's focus [**Author's note**: A charrette (charette or charet), sometimes called a design charrette, is an intense period of design or planning activity.]

Visited different areas of the base to collect information for evaluation, reviewed historical records

Documentation:

All of my work was performed within the constraints of the contractor's direction and management plan

Learning Opportunities:

Familiarity with the Defense Department's means and methods

Insight into the layout of military bases

Project Title:

Central Plant Management Improvement Plan

Summary Description:

Oversight and evaluation of all utilities and physical plant equipment

Creation of a plant improvement program for each of the buildings on the compound

Post occupancy evaluation (POE) of physical plant management, systems and equipment

Assessment and corrective actions on all facets of the operation

My Position(s):

Central Plant Manager / Assessor

Associated Duties: (Physical Plant Equipment and Systems)

Oversaw sectional replacement of cathodic protected underground 12" steam supply and condensate lines

Coordinated operations during the re-insulation of the incoming 230KV-15KV supply busses

Reviewed plans and coordinated the installation of a dual fuel, 150 psi, and 50,000-lb/hr water tube boiler

Rerouted overhead utility lines and oil storage tank lines and pumps for five 30,000-gallon USTs (underground storage tanks)

Oversaw rigger dismantling and assembly of eight 3.67 MW (megawatt) gas turbine driven generators

Provided the conceptual design and oversaw the installation of an electronically monitored cooling tower water treatment system

Supervised the laser alignment of fourteen 200 hp chilled water pumps, condenser water pumps and seven cooling tower fans

Supervised the installation of two additional 1,350-ton centrifugal chillers

Reworked the plant sanitary and storm drains

Associated Duties: (Physical Plant Management)

Determined the technical goals of the program

Worked with external stakeholders to address issues, risks and dependencies

Promoted the achievement of program objectives

Devised plans for each phase of the program objectives

Procured appropriate resources

Performed quality control checks

Evaluated engineering costs

Identified deficiencies and suggested ways to improve the delivery of service

Solicited and negotiated MEP (mechanical, electrical, plumbing) contracts

Resolved disputes and drove positive change within the program

Defined technical solutions

Performed technical planning, system integration, verification and validation

Preparation:

Review of several (smaller) prior POEs
Physical survey of the entire facility
Study of the plants historical operations
Review of equipment maintenance documents

Documentation:

Project charter
Project management plan
Equipment operating instruction manuals

Learning Opportunities:

Insight into the intricacies of government security activities
Study of existing policies and procedures
Updates on equipment and systems

Chapter 12

Project Management Essentials

Projects! There are innumerable things that can be done and countless ways of doing them, from carving a ham to building a dam, from raising a goat to floating a boat or from hanging a door to running a war. The discourse in the last chapter neither covers the whole breadth of my project undertakings, nor the full depth of my understanding of the profession, but it does show just how diverse the discipline can be. In my sampling, the diversity of the projects was obvious—citing the nature of the works, the stakeholders they involved and the documentation that was used for their execution. Projects of differing sizes, types and complexities do not lend themselves to cookie cutter approaches but, that said, there are certain elements that are common to most of them, such as lessons learned, project charters, project management plans, Gantt charts, work breakdown structures, and network diagrams. I'll do my best to provide you with a primer on each of them.

LESSONS LEARNED

One of the most important aspects of project management is what we learn from our prior successes, failures, mistakes and experiences (positive and negative) made at different points in the lifecycles of our past projects. The (value added) knowledge gained from the performance of a previous project can aid in the improved design and implementation of subsequent undertakings. The key to successful projects is learning from past experiences and actively taking those lessons learned into account in the initial planning stages of future projects. That includes the identification of the learned lessons, the actions to be taken to incorporate them, and following up to ensure that the actions taken were appropriate. Lessons should be drawn from both good ideas (that improved project efficiency) and from mistakes made that led to undesirable outcomes (for use in mitigating negative and/or reinforcing positive results). All applicable, factual and technically correct data that had an impact on project operations are compiled, formalized, and stored throughout the project's duration, then discussed. Lessons learned sessions are generally held at the conclusion of a project (during the close out period) and are usually attended by the project manager, team members and selected stakeholders. Group focus is on identifying project success and project failures, and includes recommendations to improve future project performance. Discussions generally center on what was discovered regarding the project in general (what went well and what did not go well), what was realized about its management (changes needed), and what was learned about its sponsors, customers, and aspects of communication, budgeting, procurement. Best Practice approaches to "Lessons Learned" include those shown in Table 12-1.

THE PROJECT CHARTER

After "Lessons Learned" from prior projects, the first document to be associated with any project is the "Project Charter," which establishes the existence of a project, and imparts the project manager with the authority to apply resources to it. Once a feasibility study is conducted and a project is deemed viable and capable of succeeding, the charter is issued by the project initiator or sponsor. Its purpose is to capture key information about a project. It describes what content is and is not included and enables the project principles to reach consensus on the project's scope, objectives and deliverables. It also details the resources required to attain the objectives. Considerable thought should be given to the overall purpose of the charter—who will be respon-

Table 12-1. Lessons Learned

	PRACTICE	ACTIVITIES
LESSONS LEARNED	**Record and Store Data**	Record project activities throughout the project lifecycle and store them in a central repository for later reference. Include best practices, lessons learned, and other relevant project data
	Involve All Stakeholders	Have all project participants and stakeholders participate in the lessons learned process
	Take Swift Action	Act as quickly as possible to gather information to avoid people forgetting the challenges faced during the project lifecycle
	Gather Feedback	At the conclusion of your project, compile and distribute a survey to the project team, customers, and stakeholders asking for feedback on the project's successes and failures
	Project Review Meeting	Conduct a post project assessment of the feedback to identify what lessons were learned and discuss how they can be utilized for improving the outcomes of subsequent projects
	Include All Experiences	Discuss and comment on both positive and negative experiences, including any mistakes that were made and include any corrective suggestions the participants might make
	Circulate Lessons	Document what lessons were learned and share them with the project management community
	Create An Archive	Save the conclusions as historical project data and incorporate them into your organizational database
	Un-restrict The Archive	Make lessons learned from all past projects accessible for use in all other projects

sible for creating it (normally developed by the project sponsor or a manager external to the project team), the effort needed for its preparation (the project sponsor provides background information and the project manager interviews stakeholders), and what it should contain (the purpose of the project, objectives, and outcomes). Whereas project management processes and principles may change from project to project, the elements of a project charter (framework) remain the same, irrespective of the size and type of project it describes. You can choose which elements to include in the charter (i.e., scope statements, major deliverables, assumptions, constraints, risks, change control directives, etc.) and decide how much detail to give to each. It might also include the processes for identifying and approving changes to the project scope, and specify any required resources as well as the project manager's authority to manage them. On small department-based projects someone from management is the sponsor, the department manager is the primary stakeholder, formal project approval is not required, and a short overview of the plan might include the scope, deliverables, risks, resources and project manager's authority. On medium-sized endeavors, an effort should be made to identify the stakeholders, the project's priority over other projects should be justified, and project acceptance criteria, business justification and rough estimates of the resource requirements should be addressed. Large organizational projects affecting all departments should identify internal and external stakeholders, document and justify the level of financial and human resources required, and detail the risks, benefits, and strategic impacts the project can have on the total organization.

In their simplest form, all project charters should contain the following information shown in Table 12-2.

THE PROJECT MANAGEMENT PLAN (PMP)

Benjamin Franklin has been credited with coining the phrase, "If you fail to plan, you are planning to fail," and Winston Churchill is touted as stating that, "Those who fail to learn from the past are doomed to repeat it." When it comes to project management, no truer words were ever spoken! Projects that are conducted without a map for guidance are doomed to crash and burn; only a well thought out management plan will get you to your destination. A project management plan (PMP) is a formal document that defines how a project is to be executed, monitored and controlled, stating the approach the project team will take to deliver the intended project scope. Its purpose is to provide a comprehensive baseline of who will be involved, what has to be achieved by the project, how it is to be achieved, how it will

Table 12-2. Project Charter Items

	ITEM	ACTIONS
PROJECT CHARTER	**The Project Name**	Pick a name that is short and to the point; one that captures the focus of the work and is easily recognized by all the stakeholders
	Project Description	Provide a concise and accurate overview of the project
	Executive Oversight	Name the individual (the project sponsor) having overall accountability for the project; who is primarily concerned with ensuring that the project delivers the agreed upon deliverables
	The Project Lead	Name the person (Project Manager) in overall charge of the planning and execution of a particular project
	Project Goals	Compose and list high level statements that align with your organizations business goals, that provide overall context for what the project is trying to achieve
	Project Objectives:	Compose and list lower level statements that describe the products and deliverables that the project is to deliver
	Customer Deliverable(s)	List specific reports, documents, software products, or other "end products" that are scheduled for turnover to the client during or at the conclusion of the project
	Team Members	Name the project team members and list their project titles, functions and responsibilities
	Stakeholders	Identify and list all persons that can affect or be affected by the project
	Approval Signatures	
	Project Sponsor	**Project Manager**
	Date:	**Date:**

be communicated, measured and reported. The document may include baselines (for scope, schedule, and costs); subsidiary plans (for quality, human resources, communications, risk, and procurement); and any number of additional planning strategies, such as for change management and process improvement.

Project managers are responsible for knowing and maintaining the boundaries of their projects, and they do so by creating the project management plan, following inputs from the project team and key stakeholders, before passing it on to the project sponsor or customer for its approval. The PMP should be used as a reference throughout the project for clarification of unclear areas and to ensure that the unswerving management of the project is in line with organizational policies and procedures. Considered the main communication document for the project, as a project progresses, the document will require updating with the latest available (relevant) information caused by changes made which affect the essential project information and project baselines. Small organizational project plans are generally one-page documents having a short statement of the project's desired result and the acceptance criteria. They are usually created by a department head or supervisor (as primary stakeholder), where the PM defines the project's scope and decomposes it into the major deliverables, with a member of upper management posing as the sponsor. Medium-sized projects (affecting multiple departments or done for customers or clients) identify stakeholder requirements from across the organization and possess a more detailed scope statement that also covers assumptions, constraints and the major deliverables. Larger project plans employ the use of surveys and interviews to identify internal and external stakeholder's requirements and more detailed coverage of the strategic impact of the project. **Author's note**: Not all elements are necessarily used in all projects.

THE WORK BREAKDOWN STRUCTURE

To ensure the success of any project, project managers and their teams need to know exactly what it encompasses, at a level of understanding that is straightforward and easy to digest but not overly cumbersome to monitor and control. A tool commonly applied for that purpose is the "work breakdown structure"

Table 12-3. Project Management Plan Elements

	Element	Content
PROJECT MANAGEMENT PLAN	**Approvals**	The stakeholder approval signatures
	Budget & Cost Estimates	Estimated costs and the setting of an agreed budget for the duration of the project
	Change Management	The change management process to be utilized on the project
	Communication Plan	Description of the system of communications that will be used
	Constraints	A list of any known constraints affecting the project
	Contingency Plans	Identification of project risks to address high impact risk factors
	Deliverables	The products, services, or results that the project is commissioned to produce
	Executive Summary	The key elements of the project plan
	Human Resources	The team organization, roles, responsibility and requirements
	Issues	The process to be used to manage identified project issues
	Milestones	Listing of the deliverables, due date or duration and critical dependencies
	RAM	The (Responsibility Assignment Matrix) defining who is responsible for the completion of each product
	Requirements	Listing of the space, hardware/software, and other resources needed to successfully complete the project
	Risk Management	The process to be used for managing and/or mitigating risk
	Schedule	The schedule of the projects activities
	Scope Definition	Details of the project's scope and major deliverables
	Stakeholders	The results of the project's stakeholder identification and analysis

(WBS). The WBS is a multipurpose management tool used to breakdown the work tasks of large projects into smaller, more manageable elements called work groups (or tasks) to easily communicate the work involved to execute the project, and it is used in one form or another on all but the simplest of projects.

The work breakdown schedule is the backbone of your project plan. Its primary function is to communicate clear performance expectations about the project. The project teams use the WBS to develop the project schedule, resource requirements and costs. As a data structure, it serves as a framework for planning, scheduling, scope definition, budgeting, cost estimating and contract performance measurement.

There are many ways you can present the WBS for your project. For instance, you can include a high level WBS within the project plan, then a detailed version as an appendix to the plan. If assembled correctly, every team member can view their assignments and plan how they will proceed. The tree structure view shown in Figure 12-1 is the most popular format for the work breakdown structure.

It presents an easy to understand view into the WBS; however, it is also tricky to create without an application specifically designed for creating this organizational chart structure. The outline view presents an easy to view and understand layout for the WBS. Table 12-4 illustrates my use of the tool to lay out the tasks I imposed on myself for writing this book!

Built using Microsoft Word, it was a good layout for me, as I could easily make changes, due to its auto numbering feature. Both views offer the same breakdown capability—manipulating breakdown hierarchical subdivisions of a project into work areas, with the lowest generally being a work package or sometimes even an activity. The most efficient data entry method is to start at the top of a WBS and complete each level of tasks before dropping to the next level. On large projects you will want to develop detailed levels using the 8 to 80 rule, where the WBS is broken down into work packages containing between 8 and 80 hours of work to complete. Another good rule to follow (the 1-5-5-5), suggests that each level be broken down into no more than five sublevels. Level 1 is the entire system and/or program, a program element, project or subprogram. Level 2 elements are the major elements subordinate to the Level 1 elements. Level 3 elements are subordinate to the Level 2 major elements. Representing a further defi-

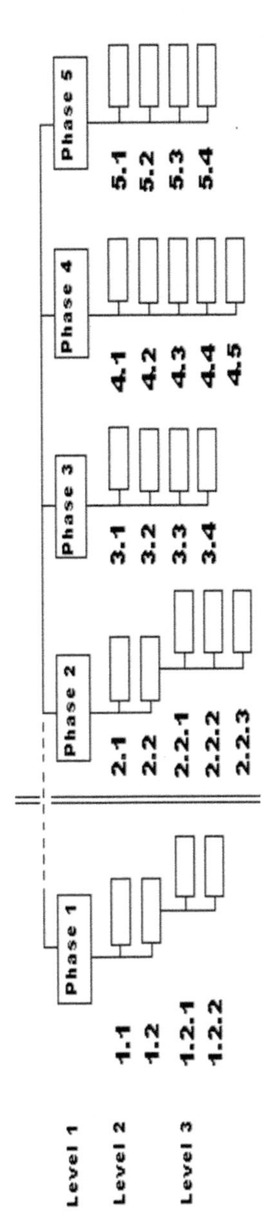

Figure 12-1. Tree Structure View of a WBS

Table 12-4. Work Breakdown (worksheet)

PROJECT MANAGEMENT BOOK (with subtitle)			
Level L	**Task Levels BS**	**Task Level Descriptions Description**	
1		**Subject Matter Research Phase**	
2	**1.1**	**Purchase Writer's Digest**	
3	**1.1.1**	Look for receptive publishers	**3 days**
--	**1.1.2**	Choose which to query	**1 day**
2	**1.2**	**Access Personal Library**	
3	**1.2.1**	Study project textbooks	**14 days**
2	**1.3**	**Access PMI Web Site**	
3	**1.3.1**	Read the PMBOK	**21 days**
4	**1.3.1.1**	Take copious notes	
2		**Contract Acquisition Phase**	
2	**2.1**	**Create a Chapter Outline**	**7 days**
2	**2.2**	**Write Query Letters**	**1 day**
3	**2.2.1**	Send queries to publishers	**1 day**
3	**2.2.2**	Evaluate responses	
3	**2.2.3**	Choose publisher	**1 day**
2	**2.3**	**Negotiate Contract**	
3	**2.3.1**	Determine submission date	
3	**2.3.2**	Sign the agreement	
3		**Manuscript Writing Phase**	
2	**3.1**	**Write Chapter Drafts**	**96 days**
3	**3.1.1**	Review and edit drafts	**72 days**
4	**3.1.1.1**	Finalize chapter narratives	**12 days**
2	**3.2**	**Insert Tables and Figures**	**2 days**
2	**3.3**	**Create & Insert Support Docs**	**14 days**
2	**3.4**	**Submit Manuscript**	

nition, Level 4 elements follow the same process of breakdown and are elements subordinate to Level 3. Level 5 elements follow the same process of breakdown and are elements subordinate to Level 4, representing an even further breakdown… and so on.

THE GANTT CHART

A Gantt chart is a graphic (bar chart) portrayal of a project, represented by horizontal lines, drawn in proportion to the duration of an activity. It shows the activities to be completed and the time needed to complete them, as well as their dependency relationships. The Gantt chart feature in today's project management software is created from the work breakdown structure, showing the project's decomposed tasks, their start and finish dates, who they are assigned to, their durations, interdependencies and often the resource or names of the resources that they are assigned to. While there are a number of free Gantt chart creators that are available for download, it's also possible to create these project management tools with software products that you may already have on your PC, most commonly using Excel or MS Project. In general, Excel is probably not the best tool for creating complex or multi-project Gantt charts, but it can be used for scheduling purposes. PMs often use Excel for small to medium-sized projects, but prefer to use Microsoft Project on more complicated undertakings. As a dedicated project management application, Microsoft Project is capable of producing much more sophisticated Gantt charts than most other tools. Ergo, it's my preference; but, in all fairness, there are a great number of project management software applications available. Using the MS Project software program, the following will take you through the process of converting a hardcopy WBS into a Gantt chart.

To start, open the software's Gantt chart view screen.
In the Task Name column, list each of your project's tasks. (See Figure 12-2.)

Note: If there is no WBS column, add one by highlighting the Task Name column and choosing WBS from the scroll-down box. This will add a WBS column to the left of the Task Name column.

(See Figure 12-3.)

Note: MS Project keeps track of WBS numbers for tasks, whether or not the column is visible.

Organize the task list into a hierarchy of summary tasks and subtasks (using the indent/outdent feature to decompose them), then add the duration (usually in days) and a start date and assigned resource to each task.

(See Figure 12-4.)

To show task dependencies, link and review your tasks and their dependencies.

Note: In Gantt charts, there are three main relationships between sequential tasks: Finish to Start (FS)—FS tasks can't start before a previous, and related, task is finished. Start to Start (SS)—SS tasks can't start until a preceding task starts. Finish to Finish (FF)—FF tasks can't end before a preceding task ends. A fourth type, Start to Finish (SF), is seldom used. (See Figure 12-5.)

A Gantt chart can show the critical path (the longest duration path through the schedule). If a task on the critical path is delayed by one day, the entire project will be delayed by one day. (See Figure 12-6.)

Note: A tracking Gantt chart can be used to show progress over time, using a percentage completed for each task and whether the project is ahead or behind against today's date. Critical path is shown by white boxes. (See Figure 12-7.)

	Notes	Task Name
1		Project Office Information Collection
2		Scope
3		Determine project office scope
4		Document requirements
5		Secure executive sponsorship
6		Scope complete
7		Planning
8		Analysis/Logistics
9		Analyze business objectives
10		Define mission statement
11		Define goals
12		Logistics complete
13		Policies and Procedures
14		Define processes
15		Define tools
16		Define techniques
17		Define compliance criteria
18		Define reporting requirements
19		Define Project Office priorities
20		Policies and procedures complete

W T F S S M T W T F S S M T W T F S

Figure 12-2. Task List Entered into MS Project

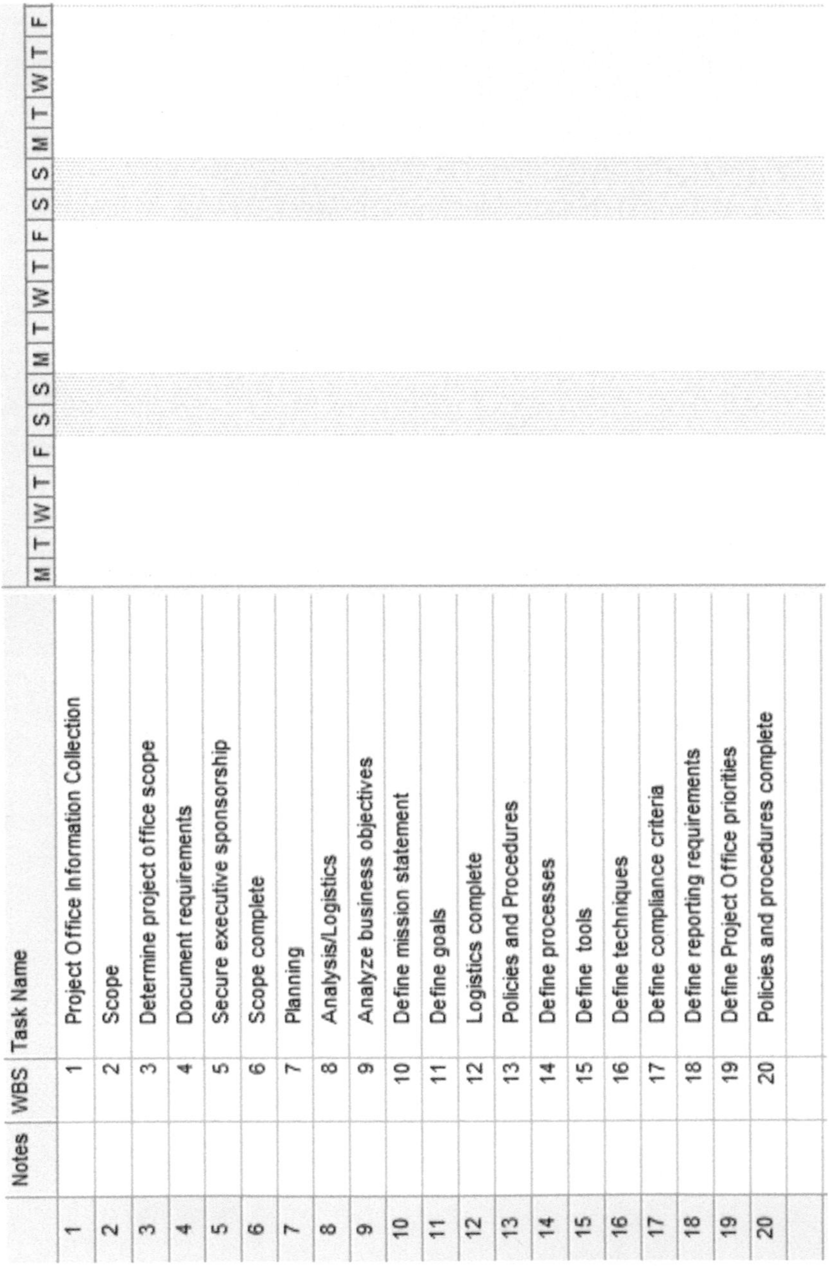

	Notes	WBS	Task Name
1		1	Project Office Information Collection
2		2	Scope
3		3	Determine project office scope
4		4	Document requirements
5		5	Secure executive sponsorship
6		6	Scope complete
7		7	Planning
8		8	Analysis/Logistics
9		9	Analyze business objectives
10		10	Define mission statement
11		11	Define goals
12		12	Logistics complete
13		13	Policies and Procedures
14		14	Define processes
15		15	Define tools
16		16	Define techniques
17		17	Define compliance criteria
18		18	Define reporting requirements
19		19	Define Project Office priorities
20		20	Policies and procedures complete

M T W T F S S M T W T F S S M T W T F

Figure 12-3. WBS Column Inserted into MS Project

	WBS	Task Name	Duration	Start	Finish	Resource Names
1	**1**	**⊟ Project Office Information Collection**	**10 days**	**Mon 1/3/00**	**Fri 1/14/00**	**Corp Mgmt**
2	**1.1**	**⊟ Scope**	**7 days**	**Mon 1/3/00**	**Tue 1/11/00**	
3	**1.1.1**	**⊟ Determine project office scope**	**6 days**	**Mon 1/3/00**	**Mon 1/10/00**	Benny
4	1.1.1.1	Document requirements	6 days	Mon 1/3/00	Mon 1/10/00	Denny
5	1.1.2	Secure executive sponsorship	1 day	Tue 1/11/00	Tue 1/11/00	**Project Manager**
6	1.1.3	Scope complete	0 days	Tue 1/11/00	Tue 1/11/00	
7	**1.2**	**⊟ Planning**	**10 days**	**Mon 1/3/00**	**Fri 1/14/00**	**Project Manager**
8	**1.2.1**	**⊟ Analysis/Logistics**	**7 days**	**Mon 1/3/00**	**Tue 1/11/00**	**Project Manager**
9	1.2.1.1	Analyze business objectives	7 days	Mon 1/3/00	Tue 1/11/00	**Project Manager**
10	**1.2.2**	**⊟ Define mission statement**	**3 days**	**Tue 1/11/00**	**Thu 1/13/00**	**Corp Mgmt**
11	1.2.2.1	Define goals	3 days	Tue 1/11/00	Thu 1/13/00	Jenny
12	1.2.3	Logistics complete	0 days	Thu 1/13/00	Thu 1/13/00	
13	**1.2.3**	**⊟ Policies and Procedures**	**10 days**	**Mon 1/3/00**	**Fri 1/14/00**	**Corp Mgmt**
14	1.2.3.1	Define processes	1 wk	Mon 1/3/00	Fri 1/7/00	Kenny
15	1.2.3.2	Define tools	1 wk	Mon 1/10/00	Fri 1/14/00	Lenny
16	1.2.3.3	Define techniques	1 wk	Mon 1/3/00	Fri 1/7/00	Jenny
17	1.2.3.4	Define compliance criteria	1 wk	Mon 1/10/00	Fri 1/14/00	Penny
18	1.2.3.5	Define reporting requirements	0.8 wks	Mon 1/3/00	Thu 1/6/00	Vinnie
19	1.2.3.6	Define Project Office priorities	1.2 wks	Fri 1/7/00	Fri 1/14/00	**Project Manager**
20	1.2.3.7	Policies and procedures complete	0 days	Fri 1/14/00	Fri 1/14/00	

Figure 12-4. MS Project with Additional Columns Inserted

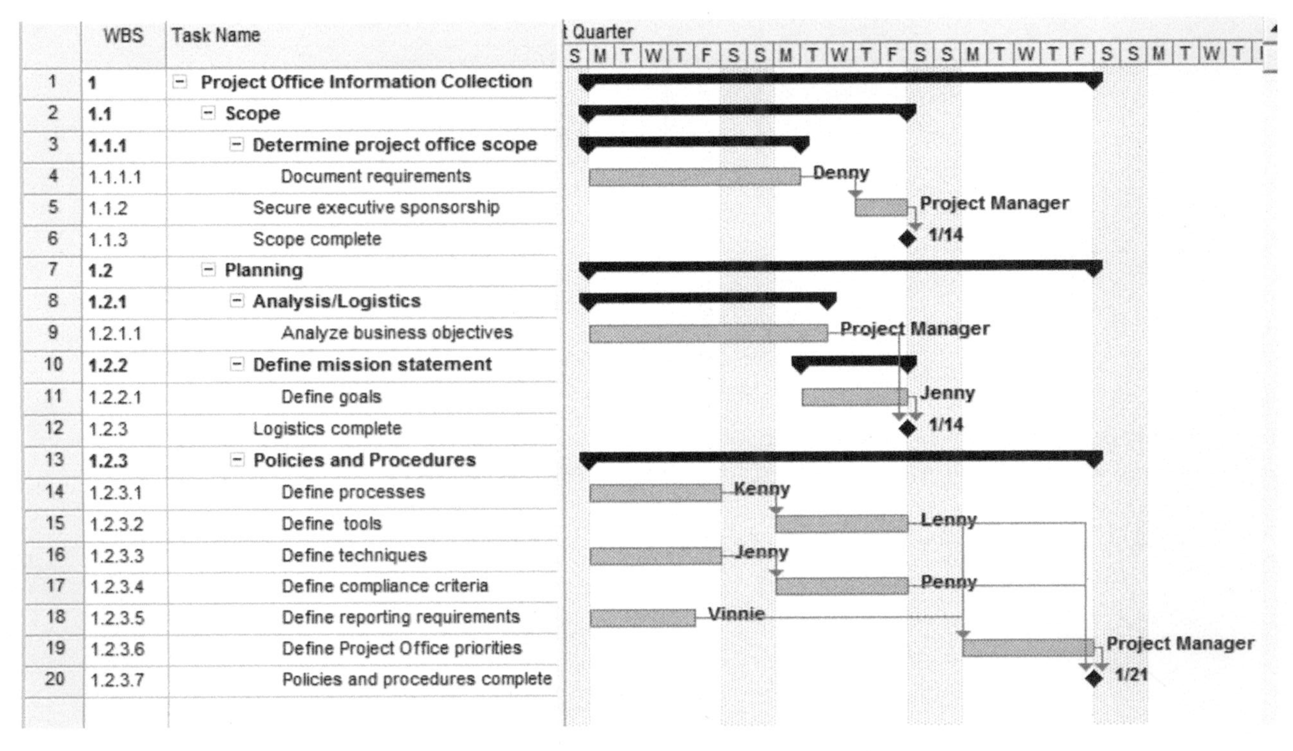

Figure 12-5. MS Project Gantt Chart Based on the WBS

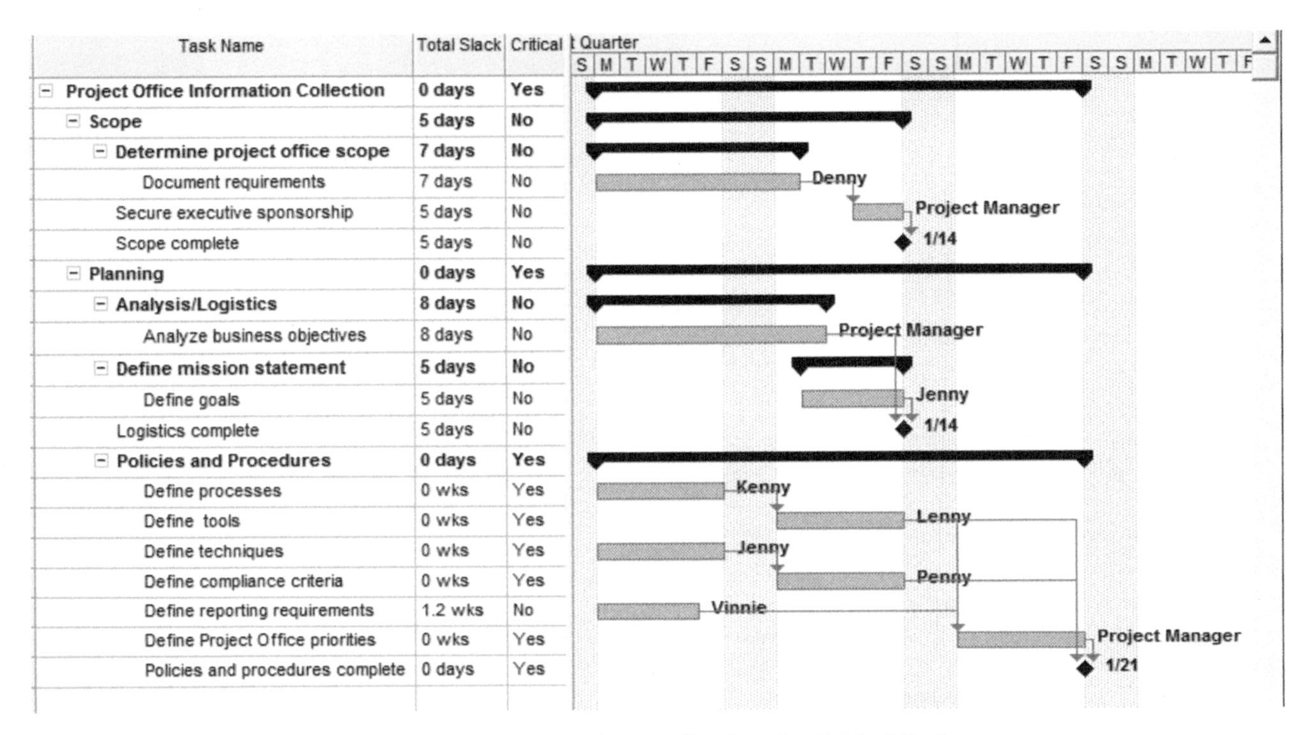

Figure 12-6. Gantt Chart Indicating the Critical Path

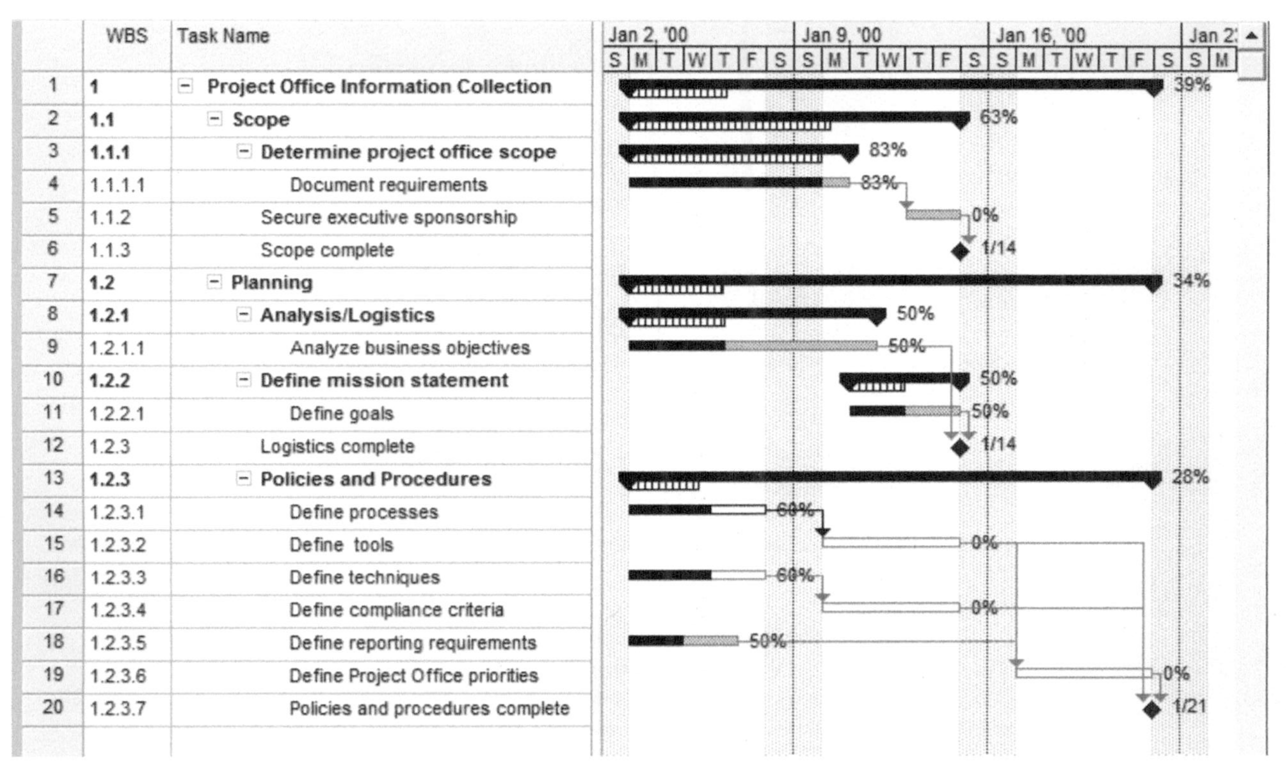

Figure 12-7. Tracking Gantt Showing the Critical Path and Percent Complete

Appendix

The Project Office

Throughout the text, we explored the singular tasks and charges of project managers but, over and above them are the duties and functions of portfolio managers. Portfolio management (by definition) is the management of a number of projects that don't share a common objective. Beyond that, are the requirements and obligations of program managers, responsible for managing multiple, ongoing inter-dependent projects. At its simpler (singular) levels, project oversight is generally provided by the project sponsor. As projects grow in size, diversity and multiplicity, a more arduous strategy is often called for. Enter the PMO (project management office).

OVERVIEW

PMOs are put into place to maximize an organization's overall business results, through the economic use of resources, across a group of projects. They afford numerous project management approaches to organizations that can differ, depending on the agency or industry, underpinning project delivery mechanisms by ensuring that changes are managed in a controlled manner. A project management office (PMO) is a group or department that defines and maintains standards for project management within an enterprise, without making any project decisions itself. It serves as the source of documentation, metrics and guidance on the practice of project management execution and functions, to provide decision support information for the organization's project, portfolio and program managers in the field. Subject to organizational undertakings, PMO teams are assembled, utilizing a mix of skill sets and roles, as needed, to ensure guidance on the execution of strategic initiatives within the organization. The PMO director (officer) role is critical, providing project enterprise-level oversight in all areas of the organization and management of resource distribution and allocation, on all projects. To accomplish that, the position is supported by numerous professional associates and administrative personnel. Successful organizations are those with enterprise PMOs that manage a staff of project managers and project support roles. Roles that complement the project manager(s) to efficiently execute programs and projects are the project planners, schedulers and controllers. They ensure that project managers are accurately informed, promoting better business decisions, allowing them to concentrate on the relevant aspects of their projects. Other staff roles include administrative backing for report generation and software support; process experts for training, quality assurance, and best practices; coordinators for researching standards, assembling project records and recording lessons learned; and resource managers to forecast the appropriate types and numbers of human and material resources. The following is information gleaned from a presentation I made to a civil engineering group, introducing them to the subject.

PMOs can be established to provide a narrow or broad set of services and include many common project management re-

A " **process** " for managing and controlling large, multi - project efforts.

A shared competency designed to integrate project management practices within an organization.

A key resource in establishing an organizational competency in project analysis, design, management and review.

11/2/2017 K L Petrocelly; PMP, IT Project+

Figure A-1. What is a PMO?

sponsibilities. PMOs often establish and deploy sets of project management processes and templates that can save project managers or organizations from having to create them on their own. They build methodologies and update them to improve best practices; facilitate improved project team communications; possess common processes, deliverables, and terminology; provide training to build core project management competencies; deliver project management coaching services; and track basic information on the current status of all projects. The PMO also assesses the general project delivery environment on an ongoing basis to determine the improvements that have been made, including actively educating and selling managers and team members on the value gained through the use of consistent project management processes.

- **Linking** project goals to an organization's strategy and business objectives
- **Synchronizing** projects to manage dependencies and gain efficiency
- **Optimizing** program scope, resources and timing
- **Ensuring** program performance exceeds expectations
- **Delivering** meaningful business value from design through implementation
- **Improving** communication and the leveraging of resources within the organization.

11/2/2017 K L Petrocelly; PMP, IT Project+

Figure A-2. PMO Objectives

Building a project management office (PMO) requires a superior knowledge and understanding of project processes. Project management skills are transferable and can be moved between applications or projects coordinated by a PMO. It is crucial to monitor and facilitate best practices and issues such as methodology and measurement, and PMOs have the knowledge and tools to consistently manage the complex coordination activities needed to implement enterprise-wide applications. A project office is a shared organizational structure that may adopt three models: project repository, project coach and enterprise PMO.

The **project repository model** is a repository of information on project management best practices and standards. This model occurs most often in organizations that empower distributed, business-centric project ownership. It's used to enfranchise the idea of consolidating or sharing management practices but falls short of direct project oversight within the organization. Project managers continue to report to, and are funded by, their respective organization areas.

The **project coach model** is a competency center to provide project expertise and some project oversight for the organization, using an internal consultancy to oversee projects, with the authority to evaluate, approve and monitor the organization's projects. The PMO serves as a source of information on project methodology and standards, and facilitates the use of a cohesive set of tools for project design, management and reporting. Project performance is actively monitored and the PMO may perform project portfolio management but usually has only input (not authority) to review and approve projects.

The **enterprise PMO** is an internal consultancy to oversee projects with more authority to evaluate, approve and monitor the organization's projects, and it is the most permanent, consolidated organizational model. It concentrates project management within the PMO and has direct management or oversight of projects wherever they occur within the organization. The PMO is involved in all projects, regardless of size, allowing it to assess scope, allocate resources, and verify time, budget and risk impact assumptions, before the project is undertaken.

Operating an established PMO is one of the top reasons that drive successful project delivery. Here's a take on the effect of managing projects with and without one.

WITHOUT A PMO	WITH A PMO
Projects are selected based on the "squeaky wheel" concept	Projects are tied to proven strategies
Poorly performing projects just keep on going, and going, and going ...	"Scope Creep" and Change Orders are minimized
Project performance is usually unknown until failure is identified	A PMO reviews projects objectively & mentors and coaches the PM's
Projects follow various, if any, methodologies; and lack a defined process	Projects follow a standard methodology, through a consistent process
Project success varies with the Project Manager chosen for the work	Project success is more consistent as best practices are used by the organization
Project Managers looked at as "necessary evils" that must be dealt with	Project Managers are recognized as change leaders
Project Managers are often afraid to ask for help in dealing with project aberrations	Project Managers build a relationship of trust with the PMO
Project Managers are chosen based on technical skill, or availability	Project Managers are chosen for their ability to deliver the project
Project Managers are essentially on their own when dealing with stakeholders	The PMO supports and helps Project Managers to negotiate with stakeholders

11/2/2017 K L Petrocelly; PMP, IT Project+

Figure A-3. Effects of Managing Projects with and without a PMO

PMO IMPLEMENTATION STRATEGIES

Strategy Drivers

PM maturity
PMO charter
Perception of value
Sponsor and management support

Evolutionary/Incremental

Lower implementation risks
Lower start-up costs
Will take longer to demonstrate ROI
More suitable if high resistance to change and low management support

Revolutionary/Wholesale

Higher implementation risks
Higher startup costs
May be able to demonstrate ROI quicker
More suitable if recognition at high level that change is imperative

PMO MATURITY LEVELS

Level 1—Supports One Project

At this first level, a PMO might support only one project at a time. Project management is performed inconsistently across the organization, and success is due to the results of competent personnel. The PMO in Level 1 provides people having the skills to fill any shortfalls that may exist in the project staff. PMO personnel work together with project personnel to solve problems and ensure proper execution of certain tasks.

Level 2—Supports Several Projects in a Program

At this first level, a PMO might support only one project at a time. Project management is performed inconsistently across the organization, and success is due to the results of competent people and unusual sacrifices. Therefore, the PMO performs short-term and quick-fix activities and augmenting, mentoring and consulting functions. Augmenting is used to fill the gaps in project team resources. The PMO in Level 1 serves in a capacity similar to a "temporary agency," providing personnel of various skills to fill any shortfalls that may exist in the project staff. PMO personnel work together with project personnel as mentors to ensure proper execution of certain tasks. The PMO can also provide consulting in the form of occasional problem-solving.

Level 3—Supports a Division/Department

At Level 3, the PMO has been established as a centralized project management entity and project management methodologies are integrated with organizational procedures. It offers an understanding of basic project management practices, defines performance management policies, and provides a clear path for improvement of those policies and procedures. The PMO develops, implements and maintains a standard system for the planning, monitoring and control of projects throughout the organization.

Level 4—Supports the Entire Organization

At this level, project management supports the business goals of the organization. The PMO assumes broader responsibilities, coordinating PM initiatives (organization-wide) and assessing the contributions of project management to the organization, directing resource use through the integrated management of projects, and establishing objectives to improve project management capabilities throughout the organization.

Level 5—Supports Business Strategy and Enterprise Resource Allocations

PMOs become the focal point for project management improvement at the highest level of maturity. They focus on continuous improvement, dissemination of best practices, and training project personnel in the latest developments. The PMO performs enterprise resource planning and integrated decision making, and promotes improved performance across all projects in the organization. In such enterprises, projects are an integral part of the business.

PMO CHARTER

In general, a PMO charter is a document granting certain specified authorities (to project, portfolio or program managers) for conducting the project management endeavors of an organization. Inputs to develop a charter can be:

Charter Scope

Business needs
Sponsor
Public vs. commercial
PM maturity

Charter Document

Mission/vision
Goals/objectives
Sponsor
Service offering
PMO governance
Key performance metrics
Funding model

PMO PHASES OF DEVELOPMENT

The success or failure of projects, programs and even organizations can be consequential of the project office used to oversee them. The idea is to develop and maintain one that serves to improve portfolio, program, and project management throughout the enterprise for which it is designed. Building a PMO is very much a project management process in and of itself. To adequately serve an enterprise, a PMO must be developed through an evolutionary process that accounts for the organization's foibles, proclivities, aspirations and constraints. Once established, that roadmap must constantly be reviewed and updated to ensure its continual improvement.

Here is an example of the phases it might be subjected to:

Rollout—Charter, roles, staffing, physical location, objectives, PM tools, initial projects review, build team, define success, have proper plans.

Operational—PMO structure delineated, PM software in use, PMO needs defined, established (communications, methodology, education, training), PMO success metrics, processes improved, agreed key stakeholders, project numbering, prioritization, historical database.

Fully Functional—Achievement driven, accountabilities for achievements, templates, template use as default, main processes automated, trend analysis mentoring, activity durations derived from historical data.

Continuous Improvement—Software tools updated, success measures evolved, meaningful project numbering, project success tracking, fewer failed projects, PM career path.

PM Glossary

Acceptance—The formal process of accepting delivery of a product or a deliverable.

Acceptance criteria—Performance requirements and essential conditions that have to be achieved before project deliverables are accepted.

Accrued costs—Earmark for the project and for which payment is due, but has not been made.

Acquisition process—This process obtains the personnel and resources necessary for project work.

Action item—Something agreed to be done by a person as a result of a discussion at a meeting and usually recorded in the minutes or log of the meeting.

Activity—A task, job, operation or process consuming time and other resources.

Activity duration—Specifies the length of time that it takes to complete an activity.

Activity list—This documents all the activities necessary to complete a project.

Activity-on-arrow (AOA)—See arrow diagramming method.

Activity-on-node (AON)—See precedence diagramming method.

Actual cost (AC)—Incurred costs charged to the project budget for which payment has been made or accrued for payment.

Actual cost of work performed (ACWP)—Total costs incurred (direct and indirect) in accomplishing work during a given time period. See also earned value.

ADM—Arrow diagramming method.

Agile—A project management methodology characterized by building products that customers really want, using short cycles of work that allow for rapid production and constant revision if necessary.

Agile project management—Agile project management draws from concepts of agile software development.

Agile software development—A set of fundamental principles about how software should be developed based on an agile way of working.

Allocation—The assigning of resources for scheduled activities in the most efficient way possible.

Arrow diagram method—One of two conventions used to represent an activity in a project.

Assumption—Factor deemed to be true during the project planning process, though proof of their validity is not available.

Authorization—The decision that triggers the allocation of funding needed to carry on the project.

Backward pass—This calculates late-start and finish dates for project activities by working backwards from the project end date.

Bar chart—A diagrammed calendar schedule of project activities' start and end dates in logical order.

Baseline—The original plan (for a project, a work package, or an activity), plus or minus approved changes.

Baseline cost—The amount of money an activity was intended to cost when the schedule was baselined.

Bottom-up estimating—The most accurate estimation technique for computing total time and cost estimates for projects by preparing individual estimates for each of a project's activities and adding them together.

Budget—Quantification of resources needed to achieve a task by a set time, within which the task owners are required to work.

A comprehensive list of revenues and expenses that represent the sum of money allocated for a project.

Business analysis—The set of tasks, knowledge, and techniques required to identify business.

Calendar unit—The smallest unit of time by which project activity durations are measured.

CCB—Change control board.

Change control board—An appointed group of stakeholders who evaluate proposed changes and decide when and whether to make them.

Change control—The process of identifying, evaluating, approving, and implementing changes to a project.

Change management—A field of management focused on organizational changes that aim to ensure efficient and prompt handling of all changes made to a project.

Change management plan—Details the change control process and ensures that all project changes are managed according to procedure.

Change request—A formal document submitted to the change control board that requests changes to the finalized project management plan.

Chart of accounts—Any numbering system used to monitor project costs by category (e.g., labor, supplies and materials). The project chart of accounts is usually based upon the corporate chart of accounts of the primary performing organization. See also code of accounts.

Closure—the formal end point of a project, either because it has been completed or because it has been terminated early.

Communications log—An ongoing documentation of communication events between stakeholders managed and collected by the project manager.

Communications management plan—This plan states who will send and receive information on aspects of the project, what details are communicated, and when communications are sent.

Conflict management—The ability to manage conflict creatively and effectively.

Constraint—A limitation on a project, including cost, human resources, time limits and quality.

Contingency—A contingency is the planned allotment of time and cost or other resources for unforeseeable elements with a project.

Contingency plan—An alternative or additional course of action planned in anticipation of the occurrence of specific risks.

Cost baseline—The sum of work package estimates, contingency reserve, and other associated costs by which project performance is assessed.

Cost control point—The point within a program at which costs are entered and controlled.

Cost estimating—Estimating the cost of the resources needed to complete project activities.

Cost management plan—This plan details how project costs will be planned, funded and controlled.

Cost overrun—A cost overrun occurs when unexpected costs cause a project's actual cost to go beyond budget.

CPM—Critical path method.

Crashing—A schedule compression technique used to speed up project work by increasing the rate at which critical path activities are completed, by adding more resources.

Critical activity—Any activity on a critical path with zero or negative float.

Critical path—In a project network diagram, the series of activities which determines the earliest completion of the project.

Critical path method (CPM)—A method used to estimate the shortest length of time needed to complete a project that includes all tasks, time estimates, task dependencies and final deliverables.

CV—Cost variance.

Decomposition—The hierarchical breaking down of project deliverables into smaller components that are easier to plan and manage.

Deliverable—A final product or component that must be provided to a client on a date and in a state that has been agreed upon by stakeholders, according to contractual stipulations.

Dependency—A logical relationship between project activities in a network diagram that determines when a dependent activity may begin.

Dummy activity—An activity of zero duration used to show a logical relationship in the arrow diagramming method.

Duration—The amount of time taken to complete an activity or task from start to finish.

Early finish date (EF)—The earliest time by which a scheduled project activity can logically finish.

Early start date (ES)—The earliest possible point in time on which the uncompleted portions of an activity (or the project) can start, based on the network logic and any schedule constraints.

Earned value (EV)—A measure of the value of completed work.

Earned value management (EVM)—A means for measuring project progress using scope, schedule, and cost figures.

End user—The person or organization that will eventually use the product of a project.

Estimation—The use of estimating techniques to reach approximations of unknown values.

Execution phase—The execution phase begins after activity approval and is the phase in which the team executes the project plan.

External constraint—A constraint from outside the project network.

Fallback plan—A plan for an alternative course of action that can be adopted to overcome the consequences of a risk, should it occur (including carrying out any advance activities that may be required to render the plan practical).

Fast tracking—Reducing the duration of a project usually by overlapping phases or activities originally planned to be done sequentially.

Finish-to-finish—In a finish-to-finish relationship, a successor activity cannot finish until a predecessor activity has finished.

Finish-to-start (FS)—In a finish-to-start relationship, a successor activity cannot start until a predecessor activity has finished.

Fixed duration—A task in which the time required for completion is fixed.

Float—The amount of time that a task in a project network can be delayed without causing a delay to subsequent tasks and or the project completion date.

Forward pass—The calculation of the early start and early finish dates for the uncompleted portions of all network activities. See also network analysis and backward pass.

Free float—The amount of time by which an activity can be postponed without affecting the early start dates of a successor activity.

FS—Finish-to-start.

Functional manager—A manager responsible for activities in a specialized department or functional group.

Gantt chart—A type of bar chart that shows all the tasks constituting a project and that illustrates a project schedule.

Go/no go—A point in a project at which it is decided whether to continue with the work.

Goal—An objective or milestone set by an individual or organization.

Handover—The formal process of transferring responsibility for and ownership of the products of a project to the operator or owner.

Henry Gantt—An American mechanical engineer and management consultant, who developed the Gantt chart in the 1910s.

Histogram—A graphic display of planned and or actual resource usage over a period of time.

Human resource management plan—A human resource management plan details the roles of and relationships between personnel working on a project, and how personnel will be managed.

Impact analysis—Assessing the merits of pursuing a particular course of action.

Initiation phase—The formal start of a new project.

Inputs—The information required to start the project management process.

Integration management plan—A document that explains integration planning and details how changes to project aspects will be managed.

Integration planning—The process of deciding how project elements will be integrated and coordinated and how changes will be addressed throughout the project management process.

Issue—An immediate problem requiring resolution.

Issue log—A document that tracks project issues, the persons responsible for resolving them and includes issue status, plans for resolution and resolution deadlines.

Key performance indicator (KPI)—A metric for measuring project success and indicating where your project stands at a given moment in time.

Kickoff meeting—The initial meeting between a project team and stakeholders where the project manager relates all the goals, plans and expectations for the team.

Lag—The minimum necessary lapse of time between the finish of one activity and the finish of an overlapping activity or the delay incurred between two specified activities.

Lag time—A necessary break or delay between activities.

Late finish date (LF)—The latest possible point in time that an activity may be completed without delaying a specified milestone.

Late start date (LS)—The latest possible point in time that an activity may begin without delaying a specified milestone.

Lead—The minimum necessary lapse of time between the start of one activity and the start of an overlapping activity.

Lead time—The amount of time an activity can be brought forward with respect to the activity it is dependent upon.

Lessons learned—The sum of knowledge gained from project work, used as references and points of interest for future projects.

Life cycle—The entire process used to build a project's deliverables.

Master project—A master project comprises a number of smaller projects, called subprojects, arranged hierarchically.

Master schedule—A summary-level schedule which identifies the major activities and key milestones.

Matrix organization—A matrix organization has different lines of reporting, representing different organizational projects or functions.

Methodology—A documented process for management of projects containing the process, definitions, and roles and responsibilities.

Milestone—An event selected for its importance in the project.

Milestone schedule—A milestone schedule details the time relationships associated with project milestones.

Mitigation—Working to reduce risk by lowering its chances of occurring or by reducing its effect if it occurs.

Most likely duration—An estimate of the most probable length of time needed to complete an activity.

Murphy's Law—"What can go wrong will go wrong."

Near critical path—A series of activities with only small amounts of total float.

Near-critical activity—A near-critical activity has only a small amount of total float, or slack time.

Negative total float—time by which the duration of an activity or path must be reduced to permit a limiting imposed date to be achieved.

Net present value (NPV)—A concept that compares the present value of a unit of currency to its inflation-adjusted possible value in the future.

Network—A pictorial presentation of project data in which the project logic is the main determinant of the placements of the activities in the drawing.

No earlier than—A restriction on an activity that indicates that it may not start or end earlier than a specified date.

No later than—A restriction on an activity that indicates that it may not start or end later than a specified date.

Node—A node is a point at which dependency lines meet, in a network diagram.

Objective—A clear, concise statement about what an activity is meant to accomplish.

Optimistic duration—An estimate of the shortest length of time needed to complete a specific activity or task.

Order of magnitude estimate—Using historical data from completed projects, it is an early, imprecise idea of the time and money required to complete a project.

Output—The end product of a process.

Overall change control—The evaluation, coordination, and management of project-related changes.

Parametric estimating—Based on historical data to establish relationships between variables, it is a technique for estimating cost and duration, such as calculating unit costs and the number of units required to complete a similar activity.

Pareto chart—A Pareto chart is a combination bar chart and line graph where the bars represent category frequencies in descending order from left to right, and the line tracks the cumulative total as a percentage.

Percent complete—The percent complete indicates the amount of work completed on an activity as a percentage of the total amount of work required.

Performance reporting—Informing stakeholders about a project's current performance and future performance forecasts.

PERT (program evaluation and review technique)—A statistical tool used to visualize a project's schedule, sequence of tasks, and even the critical path of tasks that must be completed on time for the project to meet its deadline.

Pessimistic duration—An estimate of the longest length of time needed to complete a specific activity or task.

Phase (of a project)—That part of a project during which a set of related and interlinked activities are performed.

Phase gate—A phase gate is an end-of-phase checkpoint where the project leadership reviews progress and decides whether to continue to the next phase, revisits work done in the phase, or ends the project.

Planned value (PV)—The budget assigned to the work it is meant to accomplish.

Planning—The process of thinking about the activities required to create a desired goal on some scale and the development of a course of action to pursue goals or objectives.

Planning phase—The phase of the life cycle that involves creating plans for management, control and execution of a project.

PMBOK—Project management body of knowledge.

Portfolio—A grouping of projects and programs for management convenience.

Portfolio management—The management of a number of projects that do not share a common objective.

Positive float—Positive float is defined as the amount of time that an activity's start can be delayed without affecting the project completion date.

Positive variance—The amount by which actual project performance is better than planned project performance.

Precedence diagramming method (PDM)—The process of constructing a project schedule network diagram.

Precedence network—A precedence network visually indicates relationships between project activities.

Predecessor—An activity that must be completed (or be partially completed) before a specified activity can begin.

Probability and impact matrix—A visual framework for categorizing risks based on their probability of occurrence and impact.

Problem statement—A problem statement concisely states and describes an issue that needs to be solved.

Process—An ongoing collection of activities, with inputs, outputs and the energy required to transform inputs to outputs.

Process management—The act of planning, coordinating, and overseeing processes with a view to improving outputs, reducing inputs and energy costs, and maintaining and improving efficiency and efficacy.

Procurement—The securing of goods or services.

Procurement management plan—How an organization will obtain any external resources needed for a project.

Professional development unit (PDU)—A continuing education unit that project management professionals (PMPs) take to maintain certification.

Program—A portfolio of projects selected, planned and managed in a coordinated way to achieve a set of defined objectives; a single, large or very complex project with phases managed as separate projects; or a set of otherwise unrelated projects bounded by a business cycle.

Program charter—An approved document that authorizes the use of resources for a program and connects its management with organizational objectives.

Program evaluation and review technique (PERT)—A statistical tool, used in project management, designed to analyze and represent the tasks involved in completing a given project.

Program management—The process of managing multiple ongoing inter-dependent projects.

Program management office (PMO)—The office responsible for the business and technical management of a specific contract or program.

Program manager—A program manager has formal authority to manage a program and is responsible for meeting its objectives as part of organizational project management methods.

Progress analysis—The measurement of progress against performance baselines.

Project—A temporary endeavor undertaken to create a unique product, service or result.

Project baseline—A project baseline comprises the budget and schedule allocations set during the initiation and planning phases of a project.

Project calendar—A project calendar indicates periods of time for scheduled project work.

Project charter—A document (created by a project manager and formally approved by the sponsor) that details the scope, organization and objectives of a project and authorizes the project manager's use of organizational resources.

Project communications management—A subset of project management that includes the processes required to ensure proper collection and dissemination of project information.

Project cost management—A subset of project management including resource planning, cost estimating, cost control and cost budgeting in an effort to complete the project within its approved budget.

Project integration management—A subset of project management that includes the processes required to ensure that the various elements of the project are properly coordinated.

Project life cycle—All phases or stages between a project's conception and its termination.

Project management (PM)—The application of knowledge, skills, tools and techniques to project activities, to meet or exceed stakeholder needs and expectations from a project.

Project Management Body of Knowledge (PMBOK)—A collection of project management related knowledge maintained by the Project Management Institute.

Project management office—The office or department responsible for establishing, maintaining and enforcing project delivery and management processes, procedures and standards. It provides services, support and training for project managers.

Project management plan—A plan for carrying out a project, to meet specific objectives, that is prepared by or for the project manager.

Project Management Professional (PMP)—A certified practitioner of project management. Certification is offered through the Project Management Institute.

Project management software—A type of software that includes scheduling, cost control and budget management, resource allocation, collaboration, communication, quality management and documentation or administration systems, which are used to deal with the complexity of large projects.

Project management team—The members of the project team who are directly involved in project management activities.

Project management triangle—A graphic illustration of the relationships between scope, cost and schedule.

Project manager—The person tasked with initiating, planning, executing, and closing a project, and with managing all aspects of project performance through these phases.

Project network—A graph (flow chart) depicting the sequence in which a project's terminal elements are to be completed by showing terminal elements and their dependencies.

Project network—A visual representation of the activities and dependencies involved in the successful completion of a project.

Project portfolio management (PPM)—The organization of projects and programs into a single portfolio to allow prioritization based on factors such as alignment with corporate strategy, ROI, risk, applied resource levels and technology focus.

Project scope management—A subset of project management including the processes required to ensure that the project includes all of the work required, and only the work required, to complete the project successfully.

Project scope statement—The part of a project plan that details what a project is meant to achieve and describes the deliverables expected.

Project stakeholder—Any party which may affect or be affected by a project.

Project team—A set of individuals, groups and/or organizations that are responsible to the project manager for undertaking project tasks.

Project team—The group of people who work with a project manager to execute a project plan; technical issues and strategic issues related to quality control and configuration management.

Project time management—A subset of project management including the processes required to ensure timely completion of the project.

Qualitative risk analysis—A project management technique that subjectively analyzes risk probability and impact.

Quality—A degree of excellence or the lack of it, or a property of something.

Quality assurance (QA)—A set of practices designed to monitor processes and provide confidence that result in deliverables meeting quality expectations.

Quality control (QC)—The use of standardized practices to ensure that deliverables meet stakeholder expectations.

Quality management plan—Identifies stakeholders' quality expectations and details quality assurance and quality control policies to monitor results and meet these expectations.

Quantitative risk analysis—The mathematical analysis of risk probability and impact.

RAM—Responsibility assignment matrix.

Request for proposal (RFP)—A formal invitation for expressions of interest that is extended by an organization looking to procure goods or services.

Request for quotation (RFQ)—A request asking for detailed cost estimations for specific goods or services.

Requirements management plan—A requirements management plan explains how project requirements will be defined, managed and delivered.

Residual risk—Any risks that have not been or cannot be addressed by risk mitigation or risk avoidance procedures.

Resource allocation—The assigning and scheduling of resources for project-related activities.

Resource leveling—A technique that involves amending the project schedule to keep resources below a set limit.

Resources—The people, tools, money, time, and facilities needed to complete a task.

Responsibility assignment matrix (RAM)—A document that identifies those who are responsible for project activities, accountable for ensuring that work is done, consulted on work activities, and informed on work status.

RFP—Request for proposal.

RFQ—Request for quotation.

Risk—The probability of occurrence of a specific event that affects the pursuit of objectives.

Risk acceptance—The acknowledgment of a risk without taking preemptive action against it.

Risk assessment—An activity that involves identifying and examining possible risks to a project.

Risk avoidance—The act of avoiding threats that can harm an organization, its projects or assets.

Risk management plan—A document defining how project risk analysis and management is to be implemented in the context of a particular project.

Risk management—A subset of management strategies that deals with identifying and assessing risks and acting to reduce the likelihood or impact of negative risks.

Risk mitigation—Protecting project objectives from a negative risk's impact by decreasing the probability of a negative risk occurring.

Risk register—A document used to analyze the different risks facing a project and plan the appropriate response should risks occur.

Risk sharing—Mitigation of a risk by sharing it with others, usually for some consideration.

Risk transfer—A contractual arrangement between two parties for delivery and acceptance of a product where the liability for the costs of a risk is transferred from one party to the other.

Schedule—A time sequence of activities and events representing an operating timetable for performing activities and meeting milestones.

Schedule baseline—A schedule baseline is the original project schedule by which performance is assessed.

Scope—The scope of a project constitutes everything it is supposed to accomplish to be deemed successful.

Scope baseline—The set of requirements, expectations and work packages approved as project deliverables.

Scope change management—Deals with amendments to the scope as set in the scope baseline and project management plan.

Scope creep—Uncontrolled changes in a project's scope.

Slack—Calculated time span during which an event has to occur within the logical and imposed constraints of the network, without affecting the total project duration.

Slippage—The negative variance between planned and actual activity completion dates.

SOW—Statement of work.

Sponsor—The ultimate authority over a project.

Stakeholder—Any party with an interest in the successful completion of a project.

Start-to-finish—In a start-to-finish relationship, a successor activity cannot finish until a predecessor activity has started.

Start-to-start—In a start-to-start relationship, a successor activity cannot start until a predecessor activity has started.

Steering committee—A group of people who provide high-level strategic guidance on a project.

Successor activity—A successor activity logically comes after and depends on an activity immediately preceding it.

Summary activity—A summary activity combines a set of related activities and visually represents them as a single activity.

Sunk cost—A cost that cannot be recovered once spent.

SV—Schedule variance.

System—The complete technical output of the project including technical products.

Task—A unit of work or activity needed for progress towards project goals.

Team—A team is made up of two or more people working interdependently towards a common goal and a shared reward.

TF—Total float or target finish date.

Timeline—A sequential representation of project activities.

Top-down estimating—The use of historical data from similar projects to compute time and cost estimates.

Total float (TF)—The length of time an activity can be delayed from its early start date without affecting the project end date.

Total quality management (TQM)—A common approach to implementing a quality improvement program within an organization.

Value engineering (VE)—A systematic method to improve the "value" of goods and services by using an examination of function.

Variance—A discrepancy between the actual and planned performance on a project, either in terms of schedule or cost.

Virtual team—A team comprised of people from different organizations, locations or hierarchies.

WBS—Work breakdown structure.

What-if analysis—The process of evaluating alternative strategies

Work breakdown structure (WBS)—A method that defines a project and groups the project's discrete work elements in a way that helps organize and define the total work scope of the project.

Work package—A subset of a project that can be assigned to a specific party for execution.

Workaround—A response to a negative risk event.

Zero float—A condition where there is no excess time between activities.

Bibliography

A Guide to the Project Management Body of Knowledge, copyright page, edition 2 ISBN 1-880410-12-5, and edition 3 2004 ISBN 978-1-930699-45-8, and edition 4 2008 ISBN 1-933890-51-7

Baratta, A. (2006). The triple constraint: a triple illusion. Paper presented at PMI® Global Congress 2006—North America, Seattle, WA. Newtown Square, PA: Project Management Institute.

Caccamese, A. and Bragantini, D. (2012). Beyond the iron triangle: year zero. Paper presented a t PMI® Global Congress 2012—EMEA, Marsailles, France. Newtown Square, PA: Project Management Institute.

IEEE (2011), IEEE Guide—Adoption of the Project Management Institute (PMI(R)) Standard A Guide to the Project Management Body of Knowledge (PMBOK(R) Guide)—Fourth Edition

Luyet, V; Schlaepfer, R; Parlange, MB; Buttler, B. "A framework to implement stakeholder participation in environmental projects." *J. Environ. Manag*. Elsevier. 111: 213-219. doi:10.1016/j.jenvman.2012.06.026.

"Microsoft Project 2016 Preview—What is New?" *Management Yogi*. 23 May 2015. Retrieved 10 June 2015.

Project Management Institute, *A Guide to the Project Management Body of Knowledge*—Fifth Edition, Project Management Institute Inc., 2013, Page 2.

"Project Planning Tools—Popularity Ranking." *Project Management Zone*. Retrieved 6 August 2015.

Rahim, M.A. (2002). "Toward a theory of managing organizational conflict." *The International Journal of Conflict Management*. 13: 206-235. doi:10.1108/eb022874.

Renn, O. "Stakeholder and public involvement in risk governance." *International Journal of Disaster Risk Science*. Springer. 6 (1): 8-20. doi:10.1007/s13753-015-0037-6.

Sowden, Rod; Office, Cabinet (August 30, 2011). Managing successful programmes. Stationery Office. p. 59. ISBN 9780113313273.

Thomas, K. W. (1976). Conflict and conflict management. In M.D. Dunnette (Ed.), *Handbook in industrial and organizational psychology* (pp. 889–935). Chicago: Rand McNally.

Toledo, R. (2013). Triple threat. PM Network, 27(8), 27.

Tuckman, B.W. and Jensen, M.A.C. (1977) Stages of small group development revisited. Group and Organizational Studies, 2, 419-427

What is a problem? in S. Ian Robertson, *Problem solving*, Psychology Press, 2001.

Index